MATHEMATICAL CO

AN INTRODUCTION

MATHEMATICAL COSMOLOGY

AN INTRODUCTION

PETER T. LANDSBERG

AND

DAVID A. EVANS

CLARENDON PRESS · OXFORD

Oxford University Press, Walton Street, Oxford OX2 6DP

OXFORD LONDON GLASGOW NEW YORK
TORONTO MELBOURNE WELLINGTON CAPE TOWN
IBADAN NAIROBI DAR ES SALAAM LUSAKA ADDIS ABABA
KUALA LUMPUR SINGAPORE JAKARTA HONG KONG TOKYO
DELHI BOMBAY CALCUTTA MADRAS KARACHI

© P.T. Landsberg and D.A. Evans 1977

First published in paperback, with corrections, 1979

British Library Cataloguing in Publication Data

Landsberg, Peter Theodore
  Mathematical cosmology.
  1. Cosmology - Mathematics
  I. Title   II. Evans, David A
  523.1'01'51   QB981   79-40887

  ISBN 0-19-851147-7

Printed in Great Britain
by Thomson Litho Ltd, East Kilbride, Scotland

PREFACE

Man's curiosity about the universe in which he lives can be
traced back to the beginnings of human records. The wonder
and excitement engendered by this study have been the subject
of much literature, including poetry (Lucretius, Omar Khayyam,
Shelley, and many others). In the last fifty years the subject has become a formal science in the sense that the theories of cosmology can now be confronted with an increasing
weight of observational evidence.

We feel that a subject of such philosophical and scientific interest has much to contribute to our culture and to
the education of pupils at school and students at University.
However, the difficulties of the general theory of relativity,
and of establishing the connection between theory and observation, have curtailed the use of the subject as an element in
education. This book represents an attempt to open up the
path to the frontiers of knowledge in cosmology by using only
the mathematics of the last school or early University years.
This simplification is liable to mislead the superficial
reader, but we have attempted to work largely with equations
which are in agreement with relativity, though motivated by
classical or Newtonian insight. Though many new and exciting
astrophysical results could not be discussed, we have tried
to give the background into which these discoveries can be
embedded. In this way we hope that the book will not date too
rapidly and that the reader will enjoy advancing to quantitative results by simple methods.

P.T.L.
D.A.E.

CONTENTS

LIST OF TABLES

1

INTRODUCTION

1.1. WHAT IS A STAGE II BOOK?

All areas of study can be described in ordinary language in
order to gain preliminary insights. This is as true of historical topics as it is of geographical or zoological ones.
These accounts can be exciting because they can clarify broad
areas of knowledge by rapid sweeps of the imagination. Once
one's interest is aroused in one of these topics one may want
to make a more detailed study of it. For instance, in history the documents of the time, in their appropriate language,
have to be studied, and precise dates have to be supplied.
In the case of geography the detailed curves traced out by
rivers under various conditions may be analysed. The growth
of individual species may be discussed quantitatively in
biology.

In this book we are concerned with a subject which involves the natural sciences, and again one first wants to
understand its broad outlines (Stage I). Stage II involves a
more quantitative discussion in which simple mathematical
arguments begin to appear. Stage III is the attempt to reach
the whole frontier of knowledge by studying the full topic as
currently understood by the experts. In cosmology we have a
topic of immense general and philosophical interest, since it
deals with the development of the universe as a whole. Is
space finite or infinite? How did the universe evolve into
its present state, and what is its future? How can our many
observations of stars, galaxies, pulsars, and other objects
be fitted into a coherent view of the universe? Such questions have occupied men since time immemorial, and this book,
too, deals with them. However, the subject presents aspects
which are peculiar to it.

Its first peculiarity is that interest in it is wide-
spread and is easily aroused. In fact there exist several
expositions which offer excellent and popular introductions to

it.   In broadly increasing order of technicality we mention
just a few:

> H. Bondi, *The universe at large* (London: Heinemann, 1961).
>
> F.P. Dickson, *The bowl of night* (Cambridge, Mass: M.I.T.
>   Press, 1968).
>
> S.T. Butler and H. Messel (ed.), *Man in inner and outer
>   space* (Oxford: Pergamon, 1969).
>
> J.D. North, *The measure of the universe* (Oxford: Claren-
>   don Press, 1965).
>
> D.W. Sciama, *Modern cosmology* (Cambridge University
>   Press, 1971).

Here we assume that the reader's interest in the subject is
already aroused, and that he is looking for a more quantitative understanding of the main ideas.   Even if he has not read
any of these books, he is assumed ready for stages II or III.

In order to be at the frontier of knowledge (Stage III),
however, the student needs to understand the general theory
of relativity and this presents formidable obstacles for all
those who have not attained a rather high level of mathematical education.   If they have, they may turn to the following
books, again in broadly increasing order of technicality:

> P.J.E. Peebles, *Physical cosmology* (Princeton University
>   Press, 1971).
>
> R.K. Pathria, *The theory of relativity* (Oxford: Pergamon,
>   1974).
>
> G.C. McVittie, *General relativity and cosmology*, 2nd
>   edition (London: Chapman and Hall, 1965).
>
> C.W. Misner, K.S. Thorne, J.A. Wheeler, *Gravitation* (San
>   Francisco: W.H. Freeman, 1970).
>
> S. Weinberg, *Gravitation and cosmology* (New York: Wiley,
>   1972).

The following publications deal with various ways of
gaining an understanding of relativistic cosmology without
using the full apparatus of general relativity, though none
follows the path taken in this book:

> M. Berry, *Principles of cosmology and gravitation*
>   (Cambridge University Press, 1976).
>
> C. Callan, R.H. Dicke, and P.J.E. Peeples (1965).

Cosmology and Newtonian mechanics, *Am. J. Phys.* **33**, 105.

E.R. Harrison (1965). Cosmology without relativity, *Ann. Phys.* **35**, 437.

P.J.E. Peebles (1969). Cosmology for everyphysicist, *Am. J. Phys.* **37**, 410.

It is clearly of interest to facilitate the passage from stage I to stage III by the provision of stage II expositions, and this is attempted in this book. As far as we are aware, it is the first book which attempts to make the subject systematic and quantitative without developing general relativity. We hope to give the reader a reliable introduction without going beyond the mathematics learnt at school. He should in fact be able to study research papers in some of the aspects of cosmology after perusal of this book. The expert will, however, also find new interpretations and connections in this book; but he must bear with our attempts to explain what we are doing in detail, so that it may be understandable to the beginner.

The second peculiarity of cosmology is that the connection between the theoretical models and the rapidly increasing observational material is rather subtle and needs careful explanation. We have not shirked this issue, but have devoted Chapters 10 and 11 to this facet of the subject.

In this first chapter we shall ask: what scientists are involved in our subject, and what is a rough description of the universe in space and time?

1.2. PEOPLE

The cosmologist is someone who seeks to understand the universe as a whole, in as much detail as is reasonably attainable and by scientific means. The qualifying clauses in this definition both call for further comment.

Obviously a description of even the present state of the universe which allotted one sentence to each galaxy — of which there are at least 10^{10} — would already be far too detailed for the human mind. In practice the cosmologist would be very happy with a description of the major kinds of object — stars,

quasars, galaxies — in existence. Ideally this would include
a specification of the radiation they emit, e.g. intensities,
frequencies, polarizations, angular distributions, and fluc-
tuations in these quantities. From these he might hope to
infer the masses, temperatures, velocities, etc., of these
objects. He would also want to know the way these objects
are distributed in space, and how the objects and their distribution vary with time throughout the past and future of
the universe. Much work in cosmology involves far less detailed pictures or models, in which all that is considered is
the average density of matter and radiation in the universe,
and its variation with time. As a step towards a more detailed description one would want to consider the average
densities of the various types of matter (neutrinos, electrons, protons, hydrogen atoms, helium atoms, etc.) as a function of time and the nuclear reactions by which the elements
are synthesized.

The stipulation that scientific means are used rules out
a wide range of approaches, from the purely philosophical (as
instanced by the deduction of Parmenides in the fifth century
B.C. that the universe must be perfect in form and hence
spherical and unchanging) to the fundamentalist (for instance
Archbishop Ussher's calculation that the universe was created
on 23 October 4004 B.C.). However, the range of scientific
evidence relevant to the cosmologist's search is very wide,
and he is dependent on the efforts of many colleagues. These
include the following:

The physicist, who studies the basic laws — of gravitation, light propagation, nuclear reactions, and so on — to
which all proposed models of the universe must conform.

The astronomer, who describes the present appearance
of the regions of the universe within his constantly increasing range of observation.

The astrophysicist, who studies the processes occurring in
and between astronomical objects, and describes their past and
future development. This includes the atomic processes
involved in the emission of radiation and the transport of
energy by radiation, diffusion, convection, etc., through the

material of stars. He also studies the effect of processes
such as turbulence.

The geologist, who provides information about the age and
composition of the earth, of lunar rocks and of meteorites
against which any cosmological theory can be tested. For
example, the cosmologically deduced age of the universe must
exceed that of all geological objects. Also, the observed
and cosmologically deduced abundances of certain chemical
elements must be in broad agreement.

The mathematician, who provides techniques for translat-
ing the rather general laws of physics into statements about
the range of possible models of the universe, and the observational consequences of each model.

Such brief descriptions are necessarily rough. In particular the borderlines between the various disciplines are much
less well defined than is suggested above, and one scientist
may fill different roles. There is also a constant interaction between disciplines. When the mathematical consequences
of a cosmological model are difficult to reconcile with observational data, for example, a cosmologist may propose modifications of a physical law which has so far been accepted.
His reasons may be that these modifications manifest themselves only in certain extreme conditions of temperature and
pressure which occur in cosmology, but not in terrestrial
experiments, or only in very large physical systems. Such
modifications are bound to be topics of prolonged debate among
scientists, since the cosmological reasons for proposing them
are always rather indirect. They do not have the quality of
precision which one can attribute to an experiment performed
in the laboratory. To give specific examples, Dirac [1] proposed in 1937 that the gravitational constant may depend on
time. Hoyle and Narlikar [2] suggest that the mass of elementary particles depends on the local density of matter. Such
ideas are peripheral to this book (but see Section 6.2.3 and
Appendix B), which assumes the laws of physics to be independent of position in time and space.

We can now say a little more about a cosmologist's tasks.
A cosmologist is a person who works with pencil and paper — a

theoretician. Such a person analyses the results obtained by
astrophysicists, astronomers, etc., using the laws of physics
and the manipulative techniques of the mathematician. His aim
is to create a conceptual framework which is in agreement with
as wide a range of experimental data as possible. Such a
framework is called a *model*. The model can be used by the
theoretician or other scientists to suggest additional observations which may in turn lead to a confirmation, refinement,
or rejection of the model. A simple example of a choice
between models is furnished by an argument of Eratosthenes
(*c*. 273–192 B.C.). He knew that the Sun was vertically over-
head at noon on a certain day at Syene (Egypt), and made an
angle of 7° with the vertical at noon on the same day at
Alexandria, which is about 800 km distant. This observation
is consistent with a number of models for the shape of the
Earth, two of which are (i) that the Earth is flat and that
the Sun is relatively close; and (ii) that the Earth is spherical, with a circumference of (360°/7°) × 800 km, and the Sun
very distant. Since there were other reasons for believing
the Earth to be spherical, he chose the second model and
accordingly estimated the circumference of the Earth at about
41 000 km, which is quite a good estimate by present stand-
ards. The two models are illustrated in Fig. 1.1.

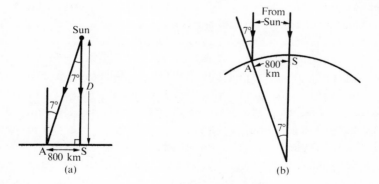

FIG. 1.1. Models of the Earth—Sun relationship available to Eratosthenes.
(a) Flat Earth, Sun at distance D = (800 cot 7°) km = 6515 km. (b) Spherical Earth, circumference of Earth = (800 × 360°/7°) km = 41143 km.

1.3. THE FRAMEWORK OF THE UNIVERSE AND ITS CONTENTS

As the astronomer directs his attention outside the solar
system, the most conspicuous objects to meet his view are
the stars. For most of history these have appeared as a constant background against which the planets move. Galileo,
using the newly invented telescope, discovered the existence
of many faint stars, invisible to the naked eye, and in particular those forming the faint band of light forming the
Milky Way. Herschel in the eighteenth century proposed that
the solar system is inside a large roughly disc-shaped system
of stars, which may be called 'our galaxy'. The Milky Way is
then simply the large number of stars in the plane of our
galaxy. It contains around 10^{11} stars, and the solar system
is about two-thirds of the way out from the centre to the
edge.

   As well as the stars, a few faint luminous patches or
nebulae are visible to the naked eye, and many more in a
telescope. Because of the disc-like appearance of some nebulae, Kant suggested that they were star systems of the same
order of size as our galaxy, but far beyond it. Other observers considered that the nebulae were all local objects. This
question was resolved in the 1920s when Hubble studied the
Andromeda nebula and estimated its distance as 800 000 light
years, putting it well outside our galaxy and suggesting that
it is comparable to it in size. Many such star systems have
since been identified. They are known as galaxies, and the
term nebula now tends to be reserved for regions of luminous
gas within galaxies. Our galaxy will be referred to as the
Galaxy with a capital G.

   A new picture of the universe had now emerged. Since
stars were not uniformly distributed in space but were clus-
tered together in galaxies, perhaps galaxies themselves should
be regarded as the basic building blocks of which the universe
is formed? Hubble estimated the distances of a large number
of galaxies, and found that they occur in clusters. However,
the average density of galaxies in space, on a large enough
scale, did not seem to vary with distance from the Earth. On
this scale, therefore, the universe appears to be uniform, and

the galaxies (or clusters) form convenient reference points in this roughly uniform framework.

Before Hubble's confirmation that the Andromeda nebula is an external galaxy, Slipher in 1912 had shown that this galaxy was approaching the Earth at a speed of about 200 km s^{-1}. The observational basis for this is a shift in the frequency of light absorbed by certain atoms in the stars of the galaxy, compared to light absorbed by atoms of the same type in the Sun or nearby stars, or in the laboratory. As discussed in Section 10.2 below, such a shift can result if the galaxy in question is approaching the Earth or receding from it. An approaching galaxy has its light shifted to higher frequency, that is, towards the blue end of the spectrum, and this is described as a blue-shift; conversely light from a receding galaxy is lowered in frequency and experiences a red-shift. Measurements on a number of galaxies followed Slipher's original observation, and in all but a few cases red-shifts were found, indicating a predominance of motions away from the Earth. Finally, in 1929, Hubble established that, except for random motions, and effects of the rotation of our own Galaxy, every galaxy observed out to a distance of six million light-years is receding from the centre of our Galaxy with a speed proportional to its distance. These galaxies are therefore continually becoming more distant from us or from the centre of our Galaxy and therefore from each other. If the galaxies are indeed the basic framework of the universe, the conclusion is clear: the universe is expanding!

The dramatic nature of the conclusion becomes evident if one looks backward in time. Each galaxy has at present a speed of recession v_0 which is proportional to its present distance r_0 from us,

$$v_0 = H_0 r_0, \qquad\qquad (1.1)$$

where H is known as the Hubble parameter and the subscript 0 indicates the present value. Observations suggest that H is roughly the same for all galaxies at the present time, although it may vary with time. This is Hubble's law.

In order to understand the meaning of the quantity H_0, let it be assumed that each galaxy has been moving away from us with its own characteristic speed which is constant in time. Then the distance of any one galaxy from us was zero at the *Hubble time* t_H given by

$$t_H \equiv \frac{1}{H_0} = \frac{r_0}{v_0}, \tag{1.2}$$

and this is the same for all galaxies. Although these equations are based on classical ideas of space and time, they continue to be valid in special and general relativity provided velocity and distance receive appropriate interpretations, as discussed in Section 10.8.

It would be wrong to infer that *all* galaxies were overlapping in some volume V of space a time t_H ago. In the first place, each galaxy might have a transverse velocity, which would, in general, depend on time. These velocities are at present believed to be small, but if they act for periods of the order t_H, they are liable to ensure that a given galaxy may never have been in the volume V. Rotating model universes are of this type. In the second place, gravitational attraction will slow down the recessional velocities of all galaxies and may do so to a different extent for different galaxies. This may mean that even those galaxies which have been in the volume V have not necessarily been there at the same time. This possibility is normally disregarded as it implies a breakdown of Hubble's law in the past.

In the simpler cosmological models of Section 6.2 it is assumed that
- (i)   the motions of the galaxies are radial with respect to us,
- (ii) the Hubble law (1.1) is valid not only now, but at all times,
- (iii) gravitation is the dominant force and acts so as to slow down the expansion.

The resulting models then indicate a time t_0 before the present at which all galaxies were at zero distance from each other. Because of (iii) the speed of recession of each galaxy was greater in the past, and therefore $t_0 < t_H$. A more realistic description would be that the galaxies formed out of less structured matter (probably hydrogen atoms), but the expansion of the universe must have started before the period of galaxy formation and can in fact be traced back to an initial very dense state, the so-called 'big bang' or 'singularity'. Because such states are not accessible to laboratory investigation, it is not certain whether currently accepted laws of physics apply to them, and it is a matter of taste how far back in time one is prepared to trust such models.   There are wide variations among scientists on this score.   It is agreed, however, that the early states of the universe were so dense that stars and planets as we know them could not exist. Thus t_0 cannot be less than the age of the Earth, estimated by geologists to be at least 4×10^9 years. Hubble's initial value for t_H was only 2×10^9 years, putting cosmologists in some difficulty.   However, Baade in 1952 showed that Hubble's estimated distances, and therefore also t_H, were wrong by a factor of about 5, due to his having wrongly identified the stars in external galaxies whose brightness he had used as an indicator of distance.   At present t_H is believed to lie in the range from 1.5×10^{10} to 2×10^{10} years, comfortably greater than the age of the Earth.

At the beginning of the 1930s, therefore, the universe was believed to consist of a large number of galaxies, distributed uniformly in space apart from local clustering, and moving apart in such a way that for every position in space a specific speed of recession is defined by eqn (1.1).   More recent observations have left this picture broadly unchanged, but have added to it, first by extending the range of vision to include more distant galaxies, and secondly by revealing new types of astronomical object, some of which can be observed at even greater distances than can ordinary galaxies. The extended range of vision is of great importance because we see distant objects by light which left them a long time ago, and are therefore observing a very ancient state of the universe.   Such observations should in principle make it possible to evaluate the merits of different cosmological models, but

in practice there are difficulties of interpretation, as discussed in Chapter 11 below. For the present we will content
ourselves with describing some of the newer types of object.

  Radio galaxies are galaxies in the normal sense, i.e.
large star systems, which emit huge amounts of energy in the
radio wavelengths. The ordinary (i.e. visible) stars in an
ordinary galaxy emit about 10^{38} erg s^{-1} in radio waves as
opposed to about 10^{44} erg s^{-1} in visible light. Radio galaxies, by contrast, emit about 10^{45} erg s^{-1} in radio waves.
The mechanism of this emission is not understood. It seems
likely that it involves catastrophic events in the centres of
galaxies, where the star densities are highest, and photo-
graphs of such galaxies often show gaseous filaments emerging
from the centre as if from a large explosion. For the cosmologist these galaxies are of interest as a class of objects
which can be detected by radio telescopes at very large distances.

  Quasi-stellar objects are also powerful radio sources,
but have an optical appearance closer to that of a visible
star than a galaxy. They also have typically very large red-
shifts, indicating by the Hubble law that they are at great
distances. This means that their total radio power output
must be very high to produce the observed radio signal
strengths. On the other hand, their physical size is believed
to be much smaller than that of a galaxy, because their radio
output can change significantly in a few years. This means
that their diameters must be at most a few light years, otherwise differences in the travel time of radio waves would smear
out the observed change in output. The combination of high
radio output and small physical size has proved very difficult
to explain, and it has been argued that quasi-stellar objects
are *not* at cosmological distances and that their red-shifts
arise either from their own gravitational fields or from high
speed motion resulting from a nearby explosion. Recent observations, however, suggest that these objects are associated
with the centres of *distant* galaxies and may in fact represent
one stage in the life of a particular type of galaxy. If so,
their distance and energy output make them of interest to the

cosmologist in the same way as radio galaxies.

Pulsars, neutron stars, and black holes may be classed
together as recently discovered objects, having in common
that they represent an extremely dense state of matter. Pulsars are radio sources which emit pulses having an extremely
constant period of theorder of 1/30 second to 1 second. The
shortness and constancy of the period mean that a pulsar must
be a small (diameter less than 1/30 light-second or 10 000 km)
and solid object. From studies of the pulsar in the Crab
nebula and others, it has been concluded that pulsars are to be
identified with neutron stars — stars which have suffered a
violent explosion, lost a great deal of their mass and energy,
and contracted to a diameter of a few tens of km. Such a star
has a density of about 10^{14} gm/cm^3. If a more massive star
runs out of its nuclear fuel and cools down, its own gravitation can compress it to even higher densities than those of
neutron stars. At a certain point in this process the
'escape velocity', that is the speed which a particle emitted
at the surface must have to leave the star entirely, may
exceed the speed of light. From then on no light ray and no
material object can leave the star, which becomes a black
hole — an entity capable of swallowing without trace any
object that approaches it too closely. Before it disappears,
matter falling into the star is greatly accelerated and com-
pressed by the star's gravitational attraction and would be
expected to emit radiation. The foregoing description is
speculative — it is claimed that black holes have been detected by their X-ray emission but other interpretations of the
data are no doubt possible. These objects can be readily
detected only if they are within our galaxy. They are clearly
of great astrophysical interest, but have less bearing on cosmology. A black hole, if at close enough range to be studied
intensively, would be of great importance to general relativity, and so indirectly to cosmology, because its intense gravitational field would give opportunities for severe tests of
this theory.

The *microwave background radiation*, though not an object
in the usual sense, is an important constituent of the uni-

verse.   Gamow [3] had suggested in 1946 that the early stages
of the universe had been hot as well as dense, and that large
amounts of radiation were present as well as matter.   In a
finite universe embedded in an infinite Euclidean space the
radiation, travelling at the speed of light, would outpace all
the galaxies and now be beyond detection.   There would, however, be no way for radiation to leave an infinite universe,
or a finite universe which fills a curved non-Euclidean space
(see Section 2.3), and the radiation must then still be present unless absorbed by matter.   Its effective temperature
would have dropped, as a result of the expansion of the universe, to a few degrees above absolute zero, but it should
still be detectable at wavelengths of a few centimetres.   Such
a radiation field was in fact discovered in 1965 [4], and
observations have been made since at a number of wavelengths.
The importance of the radiation to cosmology is twofold.
Firstly, its present temperature is one more piece of information towards understanding the early history of the universe.
Secondly, its intensity is *isotropic* to a very high degree,
that is it comes equally from all directions in space.   This
fact has been used by Collins and Hawking [5] to show that any
rotation of the universe as a whole is limited to angular
velocities less than 10^{-10} seconds of arc per century.

1.4. THE UNIVERSE IN TIME

We now turn from a rough outline of the universe in space to
an outline of its development in time.   This section introduces in a simple form some of the basic ideas discussed, in
greater detail, in Chapters 4—6 and 9.   Consider a simplified
model universe consisting of a set of n particles which move
radially in a manner defined by a universal function of time
$R(t)$, the same for all·particles, so that the position vector
of the ith particle at time t is

$$\mathbf{r}_i(t) = R(t)\, \mathbf{a}_i,   \tag{1.3}$$

$\mathbf{a}_i$ being constant for each particle.   Eqn (1.3) is equivalent
to (2.4), provided we take $R(t)$ to be a constant multiple of

the scale factor $\bar{R}(t)$ introduced there. Eqn (1.3) therefore follows from the Hubble law; it also implies the Hubble law in the sense that the speed of the ith particle is $v_i = \dot{R}(t)a_i = (\dot{R}(t)/R(t))r_i$, and v_i/r_i is the same for all particles. The kinetic energy T of the system can now be found. If m_i is the mass of the ith particle, then

$$T = \sum_{i=1}^{n} \frac{1}{2} m_i v_i^2 = \left(\sum_{i=1}^{n} \frac{1}{2} m_i a_i^2 \right) \dot{R}^2 \equiv A\dot{R}^2.$$

Also the gravitational potential energy of the system is, as discussed in Chapter 3, a sum of terms

$$- \frac{Gm_i m_j}{|\mathbf{r}_i - \mathbf{r}_j|} = - \frac{Gm_i m_j}{R(t)|\mathbf{a}_i - \mathbf{a}_j|}$$

and is therefore of the form $(-B/R)$ with B a positive constant. The total energy is

$$E = A\dot{R}^2 - B/R = A\{\dot{R}^2 - C/R\} \tag{1.4}$$

if we define $C \equiv B/A$. Detailed calculation (Section 5.1) shows that

$$C \propto \rho R^3 \tag{1.5}$$

where ρ is the density of the system.

If we consider only the n particles, the density (mass per unit volume) varies with time as R^{-3}, since the mass of any set of the particles remains constant while the volume varies as R^3. However, a realistic model universe must also contain radiation in the form of photons, and the energy of these photons contributes to the mass of the universe and so to its gravitational potential. Because of the red-shift the frequency of each photon varies with time as $1/R$ (see eqn (10.44)); the energy of each photon is Planck's constant h times its frequency and therefore also varies as $1/R$. Since the *number* density of photons varies as $1/R^3$, the energy density and the mass density of radiation will vary as $1/R^4$:

$$\rho_{(photons)} \equiv \rho_r \propto R^{-4}, \quad \rho_{(matter)} \equiv \rho_m \propto R^{-3},$$

(1.6)

$$\rho_r/\rho_m \propto R^{-1}.$$

Since the mean energy of a set of photons is a measure of its temperature T, the foregoing implies that

$$T_{(photons)} \equiv T_r \propto R^{-1}. \tag{1.7}$$

The reader may wonder why the photon density should be in the least affected by the function $R(t)$ which was defined in terms of the motion of the particles. The reason derives from the uniformity of the universe. Consider an imaginary box whose corners are located on, and move along with, a set of the particles (which we identify with galaxies). The volume of the box varies with time as R^3. Photons will cross each face of the box in both directions, but *on average*, if the universe is to remain uniform, the number entering the box must equal the number leaving. Therefore the number of photons in the box remains constant (apart from interactions with matter in the box), and the number density therefore varies as $1/R^3$. The foregoing applies to the background radiation, which has not interacted significantly with matter since very early times, rather than to radiation emitted by stars.

Eqn (1.6) points to a distinction between two stages in the history of the universe. At present the density of matter is much greater than that of radiation: $\rho \doteq \rho_m \gg \rho_r$ and so from eqns (1.5) and (1.6)

$$C \propto \rho_m R^3 = \text{constant (matter-dominated era)}. \tag{1.8}$$

As one goes back in time the ratio ρ_r/ρ_m increases — the mean energy and temperature of the background radiation increases. Eventually a time is reached before which the radiation density was much greater than that of matter: $\rho \doteq \rho_r \gg \rho_m$, and so

$$C \propto \rho_r R^3 \propto 1/R \quad \text{(radiation-dominated era)}. \tag{1.9}$$

The *matter-dominated era* covers most of the past history
of the universe and is the subject of all but Chapter 9 of
this book. In it eqn (1.4) holds with C constant, and one can
differentiate to find the rate of acceleration of the expansion:

$$0 = 2R\dot{R}\ddot{R} + C\ \dot{R}/R^2, \text{ i.e. } \ddot{R} = -C/2R^2. \qquad (1.10)$$

Evidently the expansion is slowing down. A dimensionless
measure of the rate of slowing down is the deceleration parameter q (see Section 4.6), defined by

$$q = -\ddot{R}R/\dot{R}^2 = C/2R\dot{R}^2; \qquad \dot{R}^2 = C/2Rq. \qquad (1.11)$$

Substituting this expression for $\dot{R}$ into eqn (1.4) gives an
alternative form of the energy equation:

$$E = A\{C/2Rq - C/R\} = (B/R)\{1/2q - 1\}. \qquad (1.12)$$

This result in conjunction with (1.4) is quite informa-
tive, for it shows that two main cases must be distinguished.
If $q > \frac{1}{2}$ the total energy is negative; the gravitational potential energy dominates the kinetic energy and there is enough
mass to cause the system to stop expanding and to contract.
One can see this from $\dot{R}^2 = C/R+E/A$, since if E is negative
$\dot{R}^2$ becomes zero at a finite value of R.   If $q \leqslant \frac{1}{2}$ the total
energy is positive or zero. In these cases there is not
enough mass present to force a contraction; $\dot{R}^2$ is not zero for
any finite R, and the system expands indefinitely. The future
of the universe for this simple model is essentially determined by the present value of q (see also Section 6.2), and
its observational measurement is therefore an important task
for astronomers. Some problems of observation are discussed
in Chapter 11.

The *radiation-dominated era* lasts for a time of the order
of half a million years (see Table 1.2). In it the function R
is so small that $AC/R \gg E$, and so eqns (1.4), (1.5), (1.6) imply

$$\dot{R} \doteq (C/R)^{\frac{1}{2}} \propto 1/R, \; R\dot{R} = \text{constant},$$

and, measuring time t from the instant at which R is zero,

$$R \propto \sqrt{t}, \; T_r \propto 1/R \propto 1/\sqrt{t}.$$

Thus the universe cools rapidly from initially very high temperatures when the main constituents of the universe are believed to be leptons though baryons must also be present (see Table 1.1). As it cools, more complicated structures become stable; first hydrogen and its isotopes deuterium and tritium, then helium and the heavier elements. Later galaxies and stars are formed. This is illustrated in Table 1.2.

The abundances of the elements deduced from various big-bang models can be compared with those observed in meteorites, moon rocks and on Earth, and hence act as a check on the models. Of course those elements are of special interest which are believed to have been manufactured only in the era of cosmological nucleosynthesis (and not later in stars), notably helium and deuterium. The abundance of the latter provides a particularly important constraint on the cosmological models. For if the universe was too dense at the key time, the deuterium would have been converted into helium so efficiently as to lead to too low a deuterium abundance. Hence for the big-bang nucleosynthesis to manufacture enough deuterium, it is necessary for the density of matter in the universe to be low enough. If deuterium can be made in supernovae, as has been envisaged recently [6], the constraint it provides on model universes would of course be relaxed.

The radiation from the early universe should by now have expanded to such an extent that its temperature has dropped to as low as about 3 K. This is believed to be precisely the microwave background radiation which has been detected recently (see Section 1.3).

Most of the elements in the universe are believed to have also been made in the course of thermonuclear reactions in the stars, and this *stellar* nucleosynthesis must be added to the *cosmological* nucleosynthesis. The stellar processes seem to

Particles with a continuing existence

i.e. they are constant in number.

They can be created (or destroyed) *only if* an appropriate *antiparticle* is simultaneously created (or destroyed). For example, the neutron n decays to a proton p, an electron e^- and an antineutrino $\bar{\nu}$: $n \rightarrow p + e^- + \bar{\nu}$; this conserves the number of baryons (one, before and after) and creates a lepton-antilepton pair: $e^- + \bar{\nu}$; particles are counted positive, antiparticles negative, so the total number of both leptons and baryons remains constant.

They are all *fermions*.

LEPTONS	*BARYONS*	
mass: light	mass: heavy	*QUARKS*
electron e^-	nucleons $\begin{cases} \text{proton p} \\ \text{neutron n} \end{cases}$	u,d,s,c
neutrino ν	strange baryons: Δ, Σ	Conjectured as an of the observed
muon μ^-	doubly-strange baryons: Ξ	
muon-neutrino ν_μ	triply-strange baryons: Ω^-	Baryon = 3 quarks
	charmed baryons	bound by gluons
	plus excited states	

Hadrons: particles sensitive

Can be involved in

Can be involved in the

Can be involved in the

The 1930 table would have had only the three entries in dotted boxes.

1.1

*particles**

Particles which are carriers of interactions
i.e. they are required to carry forces between particles; they *may* also
exist on their own.

Their number is not constant: they may be created (or destroyed) in
interactions.
For example, photons can be emitted (or absorbed) by atoms: the number
of photons does not remain constant.

They are all *bosons*.

	MESONS	*INTERMEDIATE VECTOR BOSONS*	*PHOTONS*	*GRAVITONS*
	(carriers of strong inter-action)	(carrier of the weak interaction)	(carrier of electro-magnetism)	(carrier of gravity)
GLUONS	mass: medium to heavy	mass: very heavy	mass: zero	
interpretation hadrons:	pion π	(W^+)	$\boxed{\gamma}$	?
	kaon	(Z°)		
	$\eta,\omega\ldots.$	(W^-)		
Meson = quark+ anti- quark	$\psi,\psi'\ldots$			
		(not yet observed)		
	plus excited states			

to the strong interaction $\longrightarrow$

the weak interaction $\longrightarrow$

electromagnetic interaction $\longrightarrow$

gravitational interaction $\longrightarrow$

*We are indebted to Dr. J.G. McEwen, Physics Department, Southampton
 University, for his help in the design of this table.

	Time (s)	Temperature (K)	Density (g/cm^{-3})	
			Radiation	Matter
Big bang	0			
Quantum era	10^{-43}			
Lepton era	2×10^{-4}	10^{12}	5×10^{13}	3×10^5
	1	10^{10}	10^5	
	24	4×10^9	2×10^3	10^{-3}
Radiation era	1000	5×10^8	1	
	10^8	10^6		
	3×10^{11}	3×10^4	10^{-17}	10^{-17}
	2×10^{13}	3000	10^{-21}	10^{-20}
Matter era	4×10^{17}	2.7	4×10^{-34}	5×10^{-30}

1 year = 3.156×10^7 s.

1.2

of the universe
rough guides only)

	Photons	Neutrinos (quarks?)

Unphysical

Not understood

Many elementary particles present, both leptons and baryons

Electron-proton pairs are still present

Neutrons, protons are still present, deuterium and tritium are beginning to be formed

---

Element synthesis (mainly $4H \rightarrow {}^4H_e$) complete, ending the period of cosmological nucleosynthesis

H, H_e, and their isotopes and small abundances of heavier elements

Time at which densities of dust and radiation are about equal

Matter and radiation decoupled

Formation of galaxies
Formation of stars
Present

2.7K   neutrinos
background   (quarks?)
radiation

bypass deuterium, lithium, beryllium, and boron. But yet
other mechanisms exist, notable spallation[†] in stellar sur-
faces or protoplanets. The existence of these processes
endows the comparison between theoretical cosmologically
deduced abundances and observed ones with some uncertainty,
even in the case of deuterium.

A rough quantitative estimate of the abundance of helium
by mass to be expected on theoretical grounds can be obtained
by the following three-step argument.

(a) Suppose nucleosynthesis commenced at a universe tempera-
     ture T_N, and that the universe then contained n neutrons
     and p protons. Because of the high temperature these
     are largely unbound. If all neutrons go to form normal
     helium (^{4}He) nuclei, $n/2$ such nuclei are formed, since
     each nucleon requires two protons and two neutrons, and
     we assume $p \gtrsim n$. The abundance in terms of mass, and
     neglecting the neutron-proton mass difference, is then

$$\frac{4(n/2)}{n + p} = \frac{2\ n/p}{1 + n/p}.$$

(b) To determine the ratio n/p, denote the neutron and proton
     masses by m_n and m_p, and assume that at temperatures of
     the order T_N they are in equilibrium according to a
     Boltzmann distribution:

$$\frac{n}{p} = \exp\left[-\frac{(m_n - m_p)c^2}{kT_N}\right] = \exp\left[\frac{-1.294}{0.862 \times 10^{-10}\ T_N}\right]$$

We have used $(m_n - m_p)c^2 \sim 1.294$ Mev, Boltzmann's constant $\sim 0.862 \times 10^{-10}$ Mev K^{-1}.

(c) To estimate T_N, observe that neutron-proton transforma-
     tions require the presence of electrons and positrons and
     their presence is therefore needed if the thermal equi-

[†]Spallation is the ejection of numerous nuclear fragments from a nucleus
when it is hit by energetic particles or other objects. This can happen,
for example, in cosmic ray processes, as a result of shock waves, in
supernovae, in nucleosynthesis, and in other processes.

librium assumption is to be valid. Their rest masses are
of order 0.511 Mev so that their pair creation is possible at universe temperatures in excess of $kT \sim 1$ Mev.
We shall take this value as E_N in $kT_N \sim E_N$ hence,

$$T_N \sim \frac{1 \text{ Mev}}{0.862 \times 10^{-10} \text{ Mev}} \sim 1.16 \times 10^{10} \text{ K}$$

$$\frac{n}{p} \sim \exp(-1.294) \sim 0.274$$

^{4}He abundance $\sim \frac{0.548}{1.274} \sim 0.43.$

The last quantity depends quite critically on the value
of E_N, as shown in Fig. 1.2. This also shows that the estimate of 43 per cent is much too high. Theories incorporating
much more data about nuclear reactions give lower estimates
closer to the empirical 25 per cent.

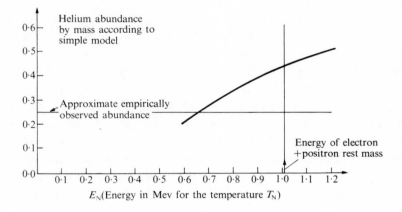

FIG. 1.2. Helium abundance by mass as a function of the typical energy
$E_N = KT_N$ of photons when nucleosynthesis is in progress.

The question of the matter density of the universe is
discussed in Section 6.2.3 and we return to the deuterium
problem briefly in Section 6.2.4. For 'stage I' discussions
of nucleosynthesis the following may be consulted:
    J.R. Gott, J.E. Gunn, D.N. Schramm, and B.M. Tinsley.
        Will the universe expand forever? *Sci. Amer.*,

March 1976.

J.M. Pasachoff and W.A. Fowler. Deuterium in the universe. ibid., May 1974.

H.-Y. Chiu. The evolution of the universe. *Science
Journal*, August 1968.

The books mentioned in Section 1.1 give more experimental
details which we have no space to discuss adequately here,
though we obtain some quantitative estimates of the temperature in equation (9.64), below.

We now turn to a preliminary discussion of various types
of models. The value of q determines the type of universe,
as has been noted above. The time-scale of the expansion is
set by the value of H. The present values of these two quantities have to be known in order to determine which of the
simplest models of the universe are good approximations to
the real universe. Of course more complicated models require
more parameters, but it is a remarkable fact that the simpler
models which require only values of H and q are surprisingly
good. These Friedmann models are studied in Chapter 5. We
include in this category models with a non-zero cosmological
constant (λ). Their description is also simple, though one
now has the value of λ as an additional parameter. The
steady-state model is also discussed at various points (e.g.
Chapter 8). Although this is an expanding model, it does not
at present furnish in a natural way two characteristics of
the big-bang type models: (a) the reasonable abundances of
deuterium, helium, and lithium (though more work is needed on
big-bang models in this connection); and (b) the microwave
background radiation. That is why the steady-state model is
not currently popular. However, in this book we seek to stand
apart from what one may regard as temporary popularity, and
concentrate on reasonably permanent features of cosmology.
The steady-state model is one of these features. Not discussed here are the anisotropic or inhomogeneous models;
though they have their own interest, they are not suitable for
a stage II discussion.

Having settled on the Friedmann models, it is clear that
the present value of q can be obtained from red-shift measure-

ments on very distant sources. These are faint and the
accuracy which can be attained is not yet very high. The
determination of the present value, H_0, of H depends on the
distances of galaxies; this is a complicated matter, discussed in Section 10.8, and there is also uncertainty in this
value.

It has already been noted that the 'Hubble time'
$H_0^{-1} \equiv t_H$ is normally larger than the time since the big bang,
so that this age must certainly be larger than the ages of
other items in the inventory of the universe. This actually
leads to the possibility of a slight overlap between the
smaller t_H-values and the larger age values of, say, the globular cluster stars $(1.4 \times 10^{10}$ years). On the whole, however,
there is no discrepancy. Other experimentally accessible
information includes determinations of the mass density (see
Section 6.2), angular diameters and number counts (see Sections 11.3 and 11.6).

1.5. THE RELATION OF COSMOLOGY TO OTHER SUBJECTS
The wide sweep of cosmology has made it a subject of interest
to people in all walks of life who seek an understanding of
the universe in which we live. It is therefore of great
philosophical importance. It has also many interactions with
other branches of science since the history of the universe
encompasses, when interpreted strictly, the geology of the
solar system, the origin of life, and the development of
man.

In connection with the theory of elementary particles, it
is believed that if quarks are their fundamental constituents,
then one would expect some to have survived from the early
universe, just as the microwave background is believed to be
a relic of the radiation of the hot early universe. Indeed
it has been estimated that the abundance of quarks now, if
they exist, should be comparable to that of gold atoms. So
one wants to know why they seem to be much rarer. What process has removed them? Or do they not exist?

In geology one is deeply interested in the question of a
possible time dependence of Newton's gravitational constant G.

This is a theme taken up quite briefly in Section 1.2 and Appendix B. In fact the general theory of relativity treats G as a constant, and the theory has to be modified to allow for variable G [7,8]. The same applies to Newtonian cosmology [9]. If G decreases with time, the radius of the Earth has increased; it has been suggested that this may be responsible for the formation of the deep basins which represent the Atlantic ocean, and there would of course be many other consequences. Most notably, the brightness of the Sun would have been much greater in the past, since it is gravitational compression which heats the cores of stars to the temperatures needed for nuclear fusion. Therefore, an increased value of G implies higher central temperature and so a greater rate of fusion. This implies further that the surface temperature of the Earth was higher in the past by an amount which, at least in the Dirac theory, may be in conflict with the fossil records [10]. The ages of the Earth, the rocks, and meteorites also represent lower limits for the time which has elapsed since the big bang.

The cosmologically deduced abundances of the elements have already been noted and observed abundances are assessed by various methods in geochemistry. There is therefore here another link between the sciences and cosmology.

The existence of life here and possibly elsewhere in the universe depends on the formation of planets and their atmospheres. This also depends on the details of cosmological evolution.

Is the universe finite or infinite? Cosmology can in principle decide this ancient question, since according to general relativity space is finite, for the simple models of section 1.4, if and only if $q > \frac{1}{2}$.

Finally, in history and philosophy the cosmological considerations of the static universe by the Greeks and in Newtonian physics have long been of interest. The cosmologies of the twentieth century throw a new light on these ideas, which enables one to reassess their underlying assumptions.

COSMOLOGY: SOME FUNDAMENTALS

2.1. THE HUBBLE LAW AND THE EXISTENCE OF A SCALE FACTOR

The implications of the galactic red-shifts observed by
Hubble in 1928, and by many others since, will now be discussed in more detail. For this purpose it will be assumed that
Hubble's law is valid at all times, not only at the present
time:

$$v(t) = H(t)r(t). \qquad (2.1)$$

This will be justified in Section 5.3.

Since the speed of recession v is the rate of change of
the distance r, one finds

$$\mathrm{d}r/\mathrm{d}t = Hr.$$

Therefore

$$\int_{r_0}^{r} \mathrm{d}r/r = \int_{t_0}^{t} H \ \mathrm{d}t,$$

The suffix 0 will normally indicate values at the present time
$t = t_0$, e.g. $H_0 = H(t_0)$. One finds

$$\ln r = \ln r_0 + \int_{t_0}^{t} H \ \mathrm{d}t$$

or

$$r(t) = r_0 \exp\left(\int_{t_0}^{t} H \ \mathrm{d}t\right) = r_0 \exp\left(-\int_{t}^{t_0} H \ \mathrm{d}t\right). \qquad (2.2)$$

This argument assumes that the Hubble law is valid at all
times between t_0 and t with H the same for all galaxies at a
given time, though possibly varying with time. Thus the exponential factor in eqn (2.2) is also the same for all galaxies.

It can be considered a universal scale factor in the sense that as one goes from time t_0 to t all intergalactic distances increase by this factor. It is usually denoted by $\bar{R}(t)$, so that for all galaxies

$$r(t) = \bar{R}(t)r_0, \quad \bar{R}(t) \equiv \exp\left(\int_{t_0}^{t} H \, dt\right). \tag{2.3}$$

Assuming no transverse motions of galaxies, the direction of the line of sight from an observer O to a galaxy remains unchanged in time. The vector equation corresponding to (2.3) is then

$$\mathbf{r}(t) = \bar{R}(t)\mathbf{r}_0 \tag{2.4}$$

where $\mathbf{r}(t)$, $\mathbf{r}_0$ are the position vectors of the galaxy under consideration, relative to O, at times t and t_0.

A note on observational aspects of the Hubble law is needed here. If a galaxy is observed by light which it emitted at a time t_0, its observed velocity v_{obs} and distance r_{obs} from the observer will be related by the Hubble parameter for time t_E: $v_{obs} = H(t_E) \, r_{obs}$. The light is observed at the later time $t_0 = t_E + r_{obs}/c$, assuming classical light propagation. Thus simultaneous observations at t_0 of galaxies at different distances r_{obs} will relate to different emission times t_E, with appropriate values of the Hubble parameter. The Hubble law (2.1) can therefore be restated in terms which make explicit allowance for the time taken by light to propagate:

$$v_{obs} = H(t_E) \, r_{obs}, \tag{2.5a}$$

$$v_{obs} = H(t_0 - r_{obs}/c)r_{obs}. \tag{2.5b}$$

(2.5a) is a general result, and the photon travel time $t_0 - t_E$ is in general relativity determined by model-dependent equations such as (11.80) to (11.82) given below with r_{obs} denoted by r. In classical theory and special relativity (2.5b) holds for all cosmological models, and is illustrated in Fig. 2.1.

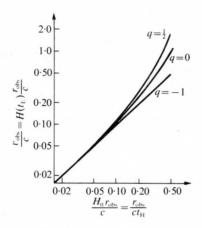

FIG. 2.1. Velocity-distance diagrams for various cosmological models
(classical and special relativistic theories of light propagation).
(a) q = -1. Steady-state model. Eqn (2.10) has been used with $H(t_E)$
treated as a constant.   (b) q = 0. Milne model based on (2.12).
(c) q = ½. Einstein-de Sitter model based on (2.18). In special relativistic cases the curves terminate at v_{obs} = c, indicating a 'horizon': see
eqn (10.30).

Note that in some popular expositions [1], curves with
q = -1 are incorectly said to refer to models with zero deceleration parameter.

Some later equations in this chapter are derived from
(2.5b) and are therefore not valid in general relativity.
Table 2.1 gives the range of validity of all equations used in
this chapter.

In Table 2.1 √ indicates equations that are valid, ×
equations that are invalid, and § correct equations obtainable
by use of a model-dependent light propagation time derivable
from (10.41).   §§ indicates that eqn (2.17) has to be replaced
by (10.32); the correct version of (2.25) is not needed here
because the Einstein-de Sitter model is not discussed in this
book for special relativity.

The first row of Table 2.1 contains results which are

TABLE 2.1

Range of validity of equations used in Chapter 2

Equation	Classical	Theories of light propagation	
		Special relativistic	General relativistic
2.1, 2, 3, 4, 5a, 6, 7, 8, 9, 10, 11, 12, 13, 16, 18, 19, 20, 21, 22	✓	✓	✓
2.5b, 14, 15, 23, 24	✓	✓	× §
2.17, 25	✓	× § §	× §

independent of light propagation, either because they do not
refer to observed quantities or, in the case of eqns (2.8) and
(2.9), because they refer to the steady-state model in which
all parameters are independent of time.

The second row contains results derived from eqn (2.5b)
and therefore not valid in general relativity. The third row
contains results derived from (2.5b) with the additional
assumption of an absolute time scale, and therefore valid
only in a completely Newtonian context.

It follows from eqn (2.5) that as one observes more and
more distant galaxies the inferred Hubble parameter changes,
since one is observing earlier states of the universe. The
simpler equation (2.1) results from a more 'god-like' view of
the universe in which one imagines the positions and velocities of galaxies to be somehow known at the same time for all
galaxies. It would be possible to distinguish in our notation
between the two types of equation, but this causes increased
complication. We have chosen instead simply to remark that
in sections dealing primarily with observed quantities, including Sections 2.4 to 2.6 and all of Chapters 10 and 11, sym-
bols such as v and r normally denote the observed values of
these quantities as in (2.5), while the discussion of cosmological models in Chapters 4 to 9 presumes the 'god-like' view
of eqn (2.1).

It also follows from eqn (2.5) that the ratio of observed
velocity to observed distance does not in general equal H_0

except in the limit of very close objects. H_0 can indeed be
defined formally as

$$H_0 \equiv \lim_{r_{obs} \to 0} \left(\frac{v_{obs}}{r_{obs}} \right) \qquad (2.6)$$

This result is valid also in special and general relativity
since the light travel-time approaches zero for nearby objects
in all these theories.

One important way of describing a cosmological model is
to give the scale factor $\bar{R}(t)$ as a function of time. This is
done, and some possible inferences from it are discussed, in
Sections 2.4 to 2.6 for three of the simpler models, after
discussion of some more general aspects of cosmology in Sections 2.2 and 2.3.

2.2. THE COSMOLOGICAL PRINCIPLE AND ITS RELATION TO THE HUBBLE LAW

At several periods in the history of cosmology it has been
thought that some object was uniquely qualified to be considered the centre of the universe. The Earth, the Sun, and our
own Galaxy have each been in this situation, and have later
been shown not to be unique — the Earth is one of several
planets, the Sun one of many stars, the Galaxy one of many
galaxies. As a result modern cosmologists have come to take
as an axiom that *no* object is in such a privileged position.
This axiom is known as the Cosmological Principle, and may be
formulated as follows:

>The large scale appearance of the universe is the same at
>a given time for all observers located in, and moving
>with, galaxies.

The phrase 'large scale' allows for possible local variations —
for instance, one galaxy may happen to be much the largest in
its immediate neighbourhood, but the principle is not violated
provided equally large or larger galaxies exist at greater
distances. The restriction to observers moving with galaxies
is necessary because an observer temporarily located in, but
not moving with, a galaxy would see other galaxies as having

transverse velocities, and so would not agree with eqn (2.4) above.

The Cosmological Principle as here formulated implies that *the universe has no unique centre*. The converse, however, is not true. One can imagine a universe which has no centre but which violates the principle — for instance, one in which the average size of galaxies increases steadily as one proceeds in a fixed direction. In fact, the available evidence supports the principle by showing that the properties of galaxies, and the number of galaxies per unit volume, do not vary with position in space out to very great distances from Earth. Thus the universe, on a large scale, appears to be *uniform* in space — there are no preferred locations. The evidence also suggests that there are no strongly preferred directions in space, i.e. that the universe is broadly *isotropic*. For instance, the Hubble parameter has roughly the same value for all galaxies, in whatever part of the sky, and the microwave background radiation comes with equal intensity from all directions in space.

The question now arises whether the Hubble law is consistent with the cosmological principle. More specifically, if the relation (2.4) is correct for an observer O, will it also be correct for observers in all other galaxies? That the answer is 'yes' can be shown as follows.

Fig. 2.2 shows the positions at times t_1 and t_2 of three observers O, A, B according to O. O considers himself to be at rest and assigns position vectors $\mathbf{r}_A(t)$ and $\mathbf{r}_B(t)$ to A and B at any time t. Applying eqn (2.3) he concludes that

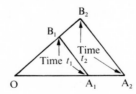

FIG. 2.2.   Positions of three observers O, A, B at times t_1 and t_2 according to O.

$$\mathbf{r}_A(t_1) = \bar{R}(t_1)\mathbf{r}_{A0}, \quad \mathbf{r}_A(t_2) = \bar{R}(t_2)\mathbf{r}_{A0},$$

$$\left.\begin{array}{c} \\ \\ \end{array}\right\} \quad (2.7)$$

$$\mathbf{r}_B(t_1) = \bar{R}(t_1)\mathbf{r}_{B0}, \quad \mathbf{r}_B(t_2) = \bar{R}(t_2)\mathbf{r}_{B0}.$$

It follows (i) that OA_1 and OA_2 are colinear, as are OB_1 and OB_2 (this has been assumed in drawing the figure), and (ii) that $OA_2/OA_1 = OB_2/OB_1 = \bar{R}(t_2)/\bar{R}(t_1)$. The triangles OA_1E_1, OA_2B_2 are therefore similar. Thus, considering the triangle formed by the three observers at different times, O will agree with the following statement:

>   **S**: In view of Hubble's law, any triangle defined by three
>   galaxies has the following properties: its angles, and
>   its orientation in space, are constant in time, while its
>   linear size is proportional to $\bar{R}(t)$.

Now the statement **S** does not single out any one corner of the triangle. Thus if one of the other observers, A for instance, applies the Hubble law based on his *own* position, he must necessarily arrive at the same statement **S**. Conversely **S** implies that each observer will see the motion of the other two as consistent with the Hubble law; that is, the *directions* of (for instance) O and B as seen from A are constant, while the *distances* are proportional to $\bar{R}(t)$. The positions from A's point of view are shown in Fig. 2.3. There is now a set of implications,

>   Hubble law for one observer $\Rightarrow$ **S** $\Rightarrow$ Hubble law for any
>   other observer, showing that the Hubble law is impartial
>   between observers, as required.

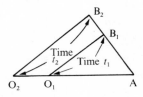

FIG. 2.3. Positions of three observers O, A, B at times t_1 and t_2 according to A.

The Hubble law also has implications for the uniformity
of the universe. Consider a set of galaxies. At any instant
the set can be considered as occupying a polyhedral volume of
space, and the location of each galaxy relative to the others
can be fixed by triangulation, i.e. by specifying the dimensions and angles of the triangles having named galaxies at
their corners. The statement **S** above will apply to each
triangle, and therefore to the entire polyhedron. Thus the
volume of the polyhedron will vary with time as $\{\bar{R}(t)\}^3$, and
so the number density (number per unit volume) of the set of
galaxies will vary as $\{\bar{R}(t)\}^{-3}$. Now $\bar{R}(t)$ is a universal function of time, independent of position in space. If, there-
fore, the number density of galaxies is uniform in space at
any one time, it will be uniform in space at all other times,
whether they be earlier or later. The only assumption made
here is Hubble's law, which enters the argument via statment
S. The situation can be summarized by saying that, if one is
only interested in the positions and velocities of galaxies,
and if the cosmological principle is known to have been true
at one instant in time, the Hubble law implies that the principle is true at all times.

It is natural to consider the converse question — does the
validity of the cosmological principle at all times imply the
Hubble law? A mathematical discussion of this is complex, but
some conclusions can be reached fairly easily in two steps.

(a) *The case of radial motion.* Consider first a universe in
which the motions of galaxies relative to an observer are
along the line of sight, but with speeds not proportional to
distance. Then eqn (1.1) is valid, but H will be different
for different galaxies; consequently eqn (2.3) is valid for
any one galaxy, but $\bar{R}(t)$ is different for different galaxies.
It follows that the volumes enclosed by different sets of
galaxies will no longer vary with time in the same manner for
all sets of galaxies. Therefore the number density of galaxies, even if it is initially uniform, cannot remain so, and
the cosmological principle must fail. The failure of Hubble's
law thus implies failure of the cosmological principle. This

is logically equivalent to the statement that the cosmological principle implies Hubble's law.

(b) *General motion*. The motion of any galaxy can in general
be regarded as the vector sum of a radial and a transverse
velocity. Now it is impossible to assign transverse velocities to every galaxy in any systematic manner without singling
out some direction in space. For instance if the vector sum
of the transverse velocities of all galaxies (within some
limiting distance) is non-zero, the direction of this vector
is evidently unique as there is a general motion of all galaxies in this direction. If, on the other hand, this vector *is*
zero, the motion of the galaxies has the general character of
a rotation, and the sum of the angular momentum vectors again
defines a unique direction. A model *can* be contrived in which
the sum of the angular momentum vectors over the whole sky is
zero, for instance by having generally clockwise motion in
two quadrants of celestial latitude and generally anticlock-
wise in the remaining two quadrants. Such a model, apart
from its extreme artificiality, singles out several directions
in space (the planes dividing the quadrants). We conclude
therefore that in a universe which is *isotropic*, i.e. has no
preferred directions in space, there can be no systematic
transverse motions of galaxies. The motions are therefore
radial and the argument of paragraph (a) above can be applied.
This gives an overall conclusion:

> *If the universe is isotropic at all times, and obeys the*
> *cosmological principle at all times, then Hubble's law is*
> *valid at all times.*

It should be noted that nothing in these arguments fixes
the *sign* of the Hubble parameter H. In fact a uniform and
isotropic universe can, at a particular time, be either expanding ($H > 0$), stationary ($H = 0$) or contracting ($H < 0$). The
evidence of the red-shifts shows that our universe is at present expanding, but in some models (see Chapters 6 and 7)
there will be a contracting phase with red-shifts replaced by
blue-shifts.

2.3. DOES THE COSMOLOGICAL PRINCIPLE REQUIRE AN INFINITE UNIVERSE?

The cosmological principle implies that the universe has no edge, since an observer at the edge would see a very non-uniform universe. This would seem to imply in Newtonian physics that the universe must be infinite, since all of infinite space must be uniformly filled with galaxies. The resulting picture of an infinite set of gravitationally interacting galaxies led to problems of the kind discussed in Section 5.3. In general relativity, by contrast, space is curved by the presence of matter, and in some models the total volume of space is finite. This can be visualized, by analogy, as follows. Consider the surface of a sphere as a two-dimensional curved 'space'. This 'space' has a finite extent so that it can be uniformly covered or 'filled' by a finite amount of two-dimensional matter. If this is done, the resulting 'universe' has no edges, and every point in it is physically identical to every other point. A being living in the surface would therefore perceive his 'universe' as satis-fying a cosmological principle, although its total 'volume' is finite.

By a slight modification the cosmological principle can be made compatible with a finite universe without resorting to curved space. One can do this by assuming that the range of vision of each observer is much smaller than the dimensions of a finite universe. Then all observers at a distance from the edge of the universe greater than the range of vision will see essentially the same picture, and these observers will far outnumber those nearer the edge. The appropriate 'modified cosmological principle' is one in which the phrase 'all observers' is replaced by 'almost all observers'. In this sense the (finite) models considered in Chapters 4 to 9 are consistent with a cosmological principle.

Before, say, 1950, when experimental evidence bearing on cosmology was still comparatively scarce and inaccurate, the cosmological principle was of great value. It was used as an important constraint on possible models. One tried to *derive* models of the universe from this and other principles. This

is no longer a desirable procedure in view of the increased
experimental evidence relevant to cosmology which is now
available. For this reason no attempt at such a derivation of
a model is made here. To put it differently, since *experimental evidence* is increasingly limiting our ignorance, we no
longer need the *cosmological principle* to do so.

2.4. THE STEADY-STATE MODEL IN NEWTONIAN COSMOLOGY

The simplest cosmological models are generally obtained by
applying some *a priori* principle to limit the variety of possible universes. The steady-state model, first proposed in
1948 [2,3] can be derived from the so-called *perfect cosmological principle* which requires that the large-scale appearance
of the universe be the same for observers at all times and at
all places. For instance, the average density of matter on
the large scale must not only be uniform in space, as required
by the cosmological principle of Section 2.2 above, but also
constant in time. Since, according to the Hubble law, all
presently existing galaxies are moving apart, new matter must
be created in the space between galaxies in order ultimately
to form new galaxies to replace those which leave any fixed
volume of space. The rate of creation needed is extremely low
on the terrestrial scale, as discussed in Section 8.2, and
would not be detectable by present experimental techniques.

The steady-state model has a number of distinctive
features. Since it has no initial dense state, it cannot be
falsified if the age of the Earth or some astronomical object
is longer than the Hubble time t_H. This can be taken to be
an advantage or disadvantage according to one's view of scientific theory in general. Also the origin of matter does not
take place only in an inaccessibly remote past but continues
at the present time where it can in principle be studied
directly. The interest of the model in the present context is
that it provides a very simple framework in which to illustrate the calculation of some observed quantities.

The results needed are straightforward. Firstly the
Hubble parameter, being a large-scale property of the universe,
must be constant in time, so that the observed Hubble law

contains H as a true constant, equal to its present value H_0:

$$v = Hr; \quad r = v/H = v/H_0 = vt_{\mathrm{H}}. \tag{2.8}$$

Secondly the number density of galaxies N_{E} must be constant in space and time, so that, if N_0 is the present density,

$$N_{\mathrm{E}} = N_0 \text{ for all } r \text{ and } t. \tag{2.9}$$

Finally, the scale factor can be calculated from eqns (2.2) and (2.3), and since H is constant we find

$$\bar{R}(t) = \exp\{H(t-t_0)\}. \tag{2.10}$$

This is a first example of a possible time dependence of a scale factor and will be used in Chapter 11 to derive rela- tions between observable quantities which can then be used to test the validity of the steady-state model.

Although the above consideration has been Newtonian, eqns (2.9) and (2.10) are valid also in special and general relati- vity since they do not depend on a theory of light propagation.

2.5. THE MILNE MODEL IN NEWTONIAN COSMOLOGY

This model, like the steady-state model, derives from an *a priori* assumption: the motions of galaxies must be such that observers in different galaxies can agree on a universal scheme of time measurement by interchanging light signals. Milne [4] showed, and we shall not attempt to prove it here, that in an expanding universe this assumption implies not only the Hubble law (2.1) but also that the velocity of each galaxy is constant in time. The main features of the model can easily be deduced from this fact.

Consider a particular galaxy which has a constant speed of recession v. The equation of motion $\mathrm{d}r/\mathrm{d}t = v$ integrates to $r = A + vt$. It is natural to choose the zero of time so that the integration constant A is zero, and then

$$r(t) = vt, \quad r_0 \equiv r(t_0) = vt_0. \tag{2.11}$$

Comparison with eqn (2.3) gives the scale factor as

$$\bar{R}(t) = r(t)/r_0 = t/t_0. \qquad (2.12)$$

The Hubble parameter is found by comparing eqn (2.11) with (2.1):

$$H(t) = v(t)/r(t) = 1/t; \; t_H \equiv 1/H_0 = t_0. \qquad (2.13)$$

Thus the Hubble parameter is infinite at the time $t = 0$, which marks the initial 'big bang' since all distances are zero at that time, and decreases with time thereafter; also the Hubble time t_H is equal to the present age of the universe t_0.

The relation between *observed* velocity v and *observed* distance r of a galaxy can be found from eqns (2.5) and (2.13):

$$v = H(t)r = \frac{r}{t} = \frac{r}{t_0 - r/c} = \frac{r}{t_H - r/c} \qquad (2.14)$$

or, re-arranging,

$$r = \frac{v t_H}{(1+v/c)}. \qquad (2.15)$$

The number density of galaxies can be calculated using the fact that, as noted in Section 2.2 above, the volume occupied by a representative set of galaxies at time t varies as $\{\bar{R}(t)\}^3$ and the number density therefore varies as $\{\bar{R}(t)\}^{-3}$ Thus if $N(t)$ is the number density at time t and $N_0 \equiv N(t_0)$ the present number density,

$$N(t) = N_0 \frac{\{\bar{R}(t_0)\}^3}{\{\bar{R}(t)\}^3} = \frac{N_0}{\{\bar{R}(t)\}^3} \qquad (2.16)$$

since by eqn (2.3) $\bar{R}(t_0) = 1$. This equation is valid for all models in which galaxies are not created (except near the start of the expansion). In the present case $R(t)$ is given by eqn (2.12), and so

$$N(t) = N_0 (t_0/t)^3.$$

Finally, if we consider only galaxies observed at distance r we have, as in eqn (2.5), that $t = t_0 - r/c$, and the corresponding number density is

$$N_E = N(t_0 - r/c) = N_0 \left(\frac{ct_0}{ct_0 - r} \right)^3 = N_0 \left(\frac{ct_H}{ct_H - r} \right)^3. \qquad (2.17)$$

Here eqn (2.13) is used to replace t_0 by t_H, since in general t_H is an observable quantity while t_0 is not.

2.6. KINEMATIC AND DYNAMIC MODELS; THE EINSTEIN—DE SITTER MODEL IN NEWTONIAN COSMOLOGY

The two models just discussed can be considered kinematic models, in the sense that they specify the *motion* of galaxies, in accordance with some *a priori* assumption, without in the first place enquiring into the forces which determine the motion. The models discussed in Chapters 5 to 9 of this book, on the other hand, are *dynamic* models, in which the forces acting on galaxies are assumed to be known and the resulting motions are discussed in terms of Newtonian or relativistic dynamics. As will appear later, a wide range of different dynamic models is consistent with our present knowledge, since such a basic quantity as the mean mass density of the universe is not known accurately. The kinematic models, on the other hand, are each completely determined once the two parameters t_H and N_0 are known. Thus the kinematic models offer the advantage of being exposed to observational disproof (for instance if the predicted dependence of v on r is not realized), but have the considerable disadvantage of not explaining the motions of galaxies in terms of currently accepted physics.

In most of the dynamic models the scale factor is a rather complex function of time, but in a much favoured model, associated with the names of Einstein and de Sitter, it is relatively simple. This model (Section 6.2.2) yields a scale factor proportional to $t^{2/3}$. The condition $\bar{R}(t_0) = 1$ therefore implies that

$$\bar{R}(t) = (t/t_0)^{2/3}. \qquad (2.18)$$

The Hubble parameter can be found as follows. If $r(t)$ and $v(t)$ are the distance and velocity of a particular galaxy at time t, then

$$r(t) = r_0 \bar{R}(t) = r_0 t^{2/3} t_0^{-2/3} \qquad (2.19)$$

and

$$v(t) = \frac{\mathrm{d}r}{\mathrm{d}t} = \frac{2}{3} r_0 t^{-1/3} t_0^{-2/3} \quad . \qquad (2.20)$$

Since by definition $H(t) = v(t)/r(t)$, eqns (2.19) and (2.20) imply

$$H(t) = \frac{2}{3t}; \quad H_0 \equiv H(t_0) = \frac{2}{3t_0} \qquad (2.21)$$

and

$$t_H \equiv \frac{1}{H_0} = 1.5 t_0, \quad t_0 = \frac{2}{3} t_H. \qquad (2.22)$$

Thus t_0, the age of the universe, is in this model two-thirds of the Hubble time t_H.

The relation between observed velocity v and observed distance r of a galaxy is again found from eqn (2.5) in conjunction with eqn (2.21):

$$v = H(t)r = \frac{2r}{3t} = \frac{2r}{3(t_0 - r/c)} = \frac{2r}{2t_H - 3r/c} \qquad (2.23)$$

or, re-arranging,

$$r = \frac{2v t_H}{2 + 3v/c} \qquad (2.24)$$

Finally the number density of galaxies at distance r is found from eqns (2.16) and 2.18):

$$N_E = N(t_0 - r/c) = N_0 \{ \bar{R}(t_0 - r/c) \}^{-3} = N_0 \left\{ \frac{t_0}{t_0 - r/c} \right\}^2$$

$$= N_0 \left\{ \frac{2c t_H}{2c t_H - 3r} \right\}^2 . \qquad (2.25)$$

A detailed discussion of some observational predictions of this and the two other models discussed is given in Chapter 11 after the necessary results of classical optics are developed in Chapter 10. For the present, Fig. 2.4 illustrates the main features of the three models.

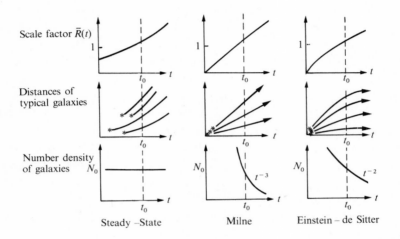

FIG. 2.4. Basic properties of some simple cosmological models. The symbol * denotes the formation of a galaxy.

PROBLEMS

(2.1) Show that, in a model universe satisfying $\bar{R}(t) = (t/t_0)^\alpha$ with α constant,

$$H = \alpha/t, \quad t_0 = \alpha t_H,$$

$$v = \frac{\alpha r}{\alpha t_H - r/c}, \quad N_E = N_0 \left\{ \frac{\alpha c t_H}{\alpha c t_H - r} \right\}^{3\alpha}$$

and that eqns (2.13, 14, 17, 21, 22, 23, 25) can be obtained on special cases of these results.

3

NEWTONIAN GRAVITATION: SOME FUNDAMENTALS

3.1. INTRODUCTION

The only important force known to act at present between
galaxies is the gravitational force of attraction. It can
therefore be expected to play a dominant role in the dynamical
behaviour of the universe. In this chapter we shall therefore
discuss gravitation. At other times in the evolution of the
universe other forces may be of comparable importance, notably
radiation pressure. These will be discussed later.

Newtonian cosmology is thus based on the principles of
classical mechanics and in particular Newtonian gravitational
theory. For our purposes it suffices, however, to assume the
inverse square law of gravitation. This states that a particle of matter of mass m_1 acts on a particle of mass m_2 at a
position $\mathbf{r} = r\hat{\mathbf{r}}$ relative to it with a force

$$\mathbf{F} = -G\frac{m_1 m_2}{r^2}\hat{\mathbf{r}}. \tag{3.1}$$

The negative sign indicates that the force tends to bring the
particles together, and $\hat{\mathbf{r}}$ is a unit vector directed from particle 1 to particle 2. The constant G is Newton's gravitational
constant:

$$G = 6.67 \times 10^{-8} \text{ dynes } \text{cm}^2\text{gm}^{-2} = 6.67 \times 10^{-8}\text{cm}^3\text{gm}^{-1}\text{s}^{-2}$$

$$= 6.67 \times 10^{-11}\text{m}^3\text{kg}^{-1}\text{s}^{-2}.$$

The last form is the value of G in SI units. The dimensions
of G are readily obtained from

$$[G] = \left[\frac{Fr^2}{m_1 m_2}\right] = \left[\frac{MLT^{-2}\cdot L^2}{M^2}\right] = [M^{-1}L^3T^{-2}].$$

It will be adequate for our discussion of Newtonian cosmology to have available one key result:

A uniform sphere of mass M and radius R attracts a particle of mass m at distance r from its centre with a force

$$F = -GmM \cdot \begin{cases} 1/r^2 & \text{(Particle external to sphere)} \\ r/R^3 & \text{(Particle inside the sphere)}. \end{cases} \tag{3.2}$$

The first part of the statement amounts to saying that the sphere may be replaced by a particle of mass M at its centre, and that the force on the particle of mass m external to the sphere is then unaltered. The second part is best written as

$$F = -GmM^*/r^2, \quad M^* \equiv (r/R)^3 M. \tag{3.3}$$

It then asserts something analogous to the first part, except that the mass at the centre of the sphere is M^*. Therefore, the only mass which is effective in producing an attraction is the mass within a sphere defined by the distance r of the mass m from the centre. It will be one object of this chapter to establish this result.

3.2. CONSERVATIVE FIELDS OF FORCE

A field of force $\mathbf{F}(r)$ is a force $\mathbf{F}(r)$ defined at all points $\mathbf{r}$ in a region of space. If a particle acted on by a force moves along a curve C drawn in this region, the field does mechanical work

$$W[\mathbf{F}(\mathbf{r}), C] = \int_C \mathbf{F}(\mathbf{r}) . d\mathbf{r}, \tag{3.4}$$

which will in general depend on the field and on the curve.

For a *conservative* field a function $U(\mathbf{r})$ exists such that

$$\mathbf{F}(\mathbf{r}) . d\mathbf{r} = -dU(\mathbf{r}) \tag{3.5}$$

The sign is such that U decreases when the particle is moving in the direction of the applied force. Also in moving from $\mathbf{r}_1$ to $\mathbf{r}_2$ the work done by the field is

$$W = -\int_{\mathbf{r}_1}^{\mathbf{r}_2} dU(\mathbf{r}) = U(\mathbf{r}_1) - U(\mathbf{r}_2). \tag{3.6}$$

The work done against the field is $U(\mathbf{r}_2)-U(\mathbf{r}_1)$. This latter quantity is also the potential energy of a particle at $\mathbf{r}_2$ relative to $\mathbf{r}_1$. We note: (a) The work done clearly depends only on the end points and not on the path. (b) In going from $\mathbf{r}_1$ to $\mathbf{r}_2$ under the action of the field, the work W is positive if $U(\mathbf{r}_1)>U(\mathbf{r}_2)$, and a particle tends to slide down a potential energy hill under the action of the field, as shown in Fig. 3.1.

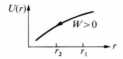

FIG. 3.1. When a particle is moved by a field of force to a point of lower potential energy, then the work done by the field is positive.

The term 'conservative' refers to the possibility of defining a function which is constant in time during the motion of a particle. This is in fact the total energy of the particle, which is thus 'conserved'. One proves it as follows:

If $\mathbf{v}$ is the velocity of the particle, then

$$\mathbf{F}\cdot\mathbf{v} = m\frac{d\mathbf{v}}{dt}\cdot\mathbf{v} = \frac{1}{2}m\frac{d\mathbf{v}^2}{dt} = \frac{d}{dt}(\frac{1}{2}mv^2), \qquad (3.7)$$

which is the rate of change of the kinetic energy of a particle. It follows that, if t is the time, then using eqns (3.5) and (3.7),

$$\int_1^2 \mathbf{F}\cdot\mathbf{v}\ dt = \int_1^2 \mathbf{F}\cdot d\mathbf{r} = U(r_1)-U(r_2) = \frac{1}{2}m(v_2^2-v_1^2).$$

Energy conservation follows in the form

$$\frac{1}{2}mv_1^2+U(\mathbf{r}_1) = \frac{1}{2}mv_2^2+U(\mathbf{r}_2).$$

As an example of a *non-conservative field of force*, consider

$$F(r) = y^2\mathbf{i}+x^2\mathbf{j}$$

where $\mathbf{i}$, $\mathbf{j}$ are unit vectors parallel to the x and y axes. Suppose this field does work on a particle as follows:

> Starting point $\mathbf{r}$ = $(0,0)$.
> Curve traced out $C: y = 2x^n$ (n is a constant)
> Endpoint: $\mathbf{r}$ = $(1,2)$.

In this case

$$\mathbf{F}(\mathbf{r}).d\mathbf{r} = (y^2, x^2).(dx, dy) = y^2 dx + x^2 dy = 4x^{2n} dx + 2nx^{n+1} dx.$$

Integrating as in eqn (3.4) yields

$$W = \left[\frac{4}{2n+1} x^{2n+1} + \frac{2n}{n+2} x^{n+2} \right]_0^1 = \frac{2(2n^2+3n+4)}{(n+2)(2n+1)}.$$

We see that this yields 18/9 for $n = 1$ and 9/5 for $n = 2$, so that the work does depend on the path, and the field is not conservative.

We now adopt the inverse square law of force (3.1) to find (see Fig. 3.2)

$$\mathbf{F}(\mathbf{r}).d\mathbf{r} = -\frac{Gm_1 m_2}{r^2}(\hat{\mathbf{r}}.d\mathbf{r}).$$

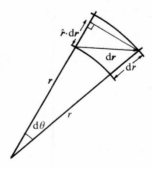

FIG. 3.2. The work done by a field of force $\mathbf{F}(r)$ involves the scalar product $\hat{\mathbf{r}}.d\mathbf{r}$.

Thus only the projection of $d\mathbf{r}$ on $\hat{\mathbf{r}}$ enters the expression for the work, so that for any path only the projection on $\hat{\mathbf{r}}$ is relevant. Thus in the limit $d\mathbf{r} \to 0$, $\mathbf{F}(\mathbf{r}).d\mathbf{r} = -Gm_1m_2 dr/r^2$ (Fig. 3.2). In that limit, therefore,

$$U(\mathbf{r}_1) - U(\mathbf{r}_2) = \int_{\mathbf{r}_1}^{\mathbf{r}_2} \mathbf{F}(\mathbf{r}).d\mathbf{r} = -Gm_1m_2 \int_{r_1}^{r_2} r^{-2} dr = -\frac{Gm_1m_2}{r_1} + \frac{Gm_1m_2}{r_2}.$$

Therefore $\mathbf{F}$ is conservative and the appropriate function U may be defined by

$$U(r) = U(\infty) - Gm_1m_2/r. \tag{3.8}$$

$U(\infty)$ is usually taken to be zero, and this gives the usual expression for the gravitational potential energy (see Fig. 3.3)

$$U(r) = -Gm_1m_2/r. \tag{3.9}$$

This expression agrees with intuition: as the particles are separated one does work against the field (their mutual attraction), and if they are allowed to move, they will collide and attain their state of least potential energy, which is $-\infty$ in this case.

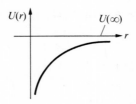

FIG. 3.3. If the gravitational potential energy $U(r)$ of a particle 2 at a distance r from a particle 1 is zero at $r = \infty$, then it is negative at all finite values of r.

3.3. THE GRAVITATIONAL POTENTIAL

The potential energy (3.9) of particle 2 in the field of particle 1 depends on the mass of particle 2. Any other particle would be acted on by an analogous force. This makes

it convenient to speak of the gravitational potential $V(r)$ due
to particle 1:

$$V(r) = V(\infty) - Gm_1/r.$$

The potential energy of *any particle of mass m* at distance r
from particle 1 is then

$$U(r) = V(r)m. \qquad (3.10)$$

Thus the potential energy is associated with at least two
particles, while the potential can be associated with one
particle only.

　　As an example, suppose a gravitating spherically symmetrical body is generated by the spherically symmetrical condensation of particles from infinity. Suppose that at some given
stage the body has a radius x. The gravitational potential at
all points of its surface has then one and the same value
which will be denoted by $V(x)$. Imagine that a new thin layer
of mass $dM(x)$ condenses from infinity on the surface. Its
potential energy at infinity is zero, and its potential energy
in the new position is $V(x)dM(x)$. The change in the potential
energy of the system as a result of this new condensation is
the same as the work done against the field, which by eqn (3.6)
is

$$dW = dU(x) - dU(\infty) = dU(x) = V(x)dM(x) \qquad (3.11)$$

This result will be useful later.

　　Another important example is discussed next.

3.4. THE POTENTIAL DUE TO A UNIFORM THIN SPHERICAL SHELL

　　The shell is divided into annuli so that all the mass
points of a typical annulus A are at a fixed distance z from
P, as shown in Fig. 3.4. Let the mass of the whole shell be
m_{sh} so that the mass of A can be denoted by dm_{sh}. The incremental width, ds say, of A is, because of the rounded nature
of the surface, larger than the projection, dx say, of the

width.   In fact (Fig. 3.4), $dx = -ds \sin \alpha$, where 2α is the angle subtended by A at the centre 0 of the shell.   This can be seen from Fig. 3.4 or algebraically from

$$x = a \cos \alpha, \quad dx = -a \sin \alpha \, d\alpha = -ds \sin \alpha.$$

Here a is the radius of the shell.   From Fig. 3.4

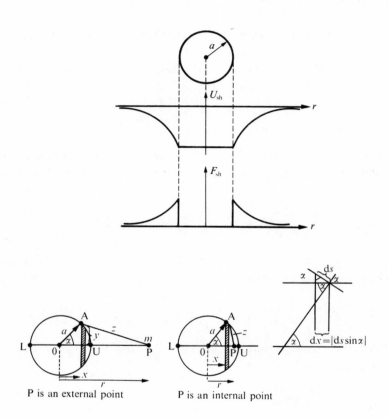

P is an external point   P is an internal point

FIG. 3.4.   The calculation of the gravitational potential energy of a particle at P due to a uniform thin spherical shell of matter of radius a and centred at 0.

$$z^2 = a^2 + r^2 - 2ar \cos \alpha = a^2 + r^2 - 2rx$$

and also

$$dx = -(z/r)dz.$$

The surface area of A is

$$2\pi(a \sin \alpha)ds = -2\pi a dx = +(2\pi a/r)z dz.$$

If we multiply this quantity by the mass per unit area, $m_{sh}/4\pi a^2$, we obtain the mass of the annulus:

$$dm_{sh} = \frac{2\pi a}{r} \cdot \frac{m_{sh}}{4\pi a^2} z dz = \frac{m_{sh}}{2ar} z dz.$$

The potential energy due to the attraction between a mass m at P and the mass of A is, after integration over all annuli A,

$$U_{sh} = -\int G \frac{m dm_{sh}}{z} = -\frac{Gmm_{sh}}{2ar} \int dz = -\frac{Gmm_{sh}}{2ar} \left. |z| \right|_L^U$$

where U, L are the upper and lower limits for z. These depend on whether P is an external or an internal point, and are

$$\left. |z| \right|_L^U = \begin{cases} (r+a)-(r-a) = 2a \text{ for an external point P} \\ (r+a)-(a-r) = 2r \text{ for an internal point P} \end{cases}$$

It follows that the potential energy U_{sh} and the force of attraction E_{sh} are

$$U_{sh} = -\frac{Gmm_{sh}}{r} \left. \right\} \text{(external point P)} \qquad -\frac{Gmm_{sh}}{a} \left. \right\} \text{(internal point P)}$$

$$E_{sh} = -\frac{Gmm_{sh}}{r^2} \qquad\qquad\qquad 0$$

(3.12)

where $F_{sh} = -dU_{sh}/dr$ (eqn (3.5) has been used. A very similar result holds for electrostatic charges, as will now be shown.

Consider a thin spherical shell of centre 0 and radius a with constant charge density $\sigma = q_{sh}/4\pi a^2$, where q_{sh} is the total charge on the shell. Let q be a point charge of opposite sign placed at a point P such that $OP = r$. Then the potential energy U_{sh} of the system and the force of attraction F_{sh} between charge and shell, are

$$U_{sh} = -\frac{qq_{sh}}{\varepsilon r} \left.\begin{array}{c} \\ \\ \end{array}\right\} \begin{array}{c} \text{(external} \\ \text{point P)} \end{array} \quad \text{or} \quad -\frac{qq_{sh}}{\varepsilon a} \left.\begin{array}{c} \\ \\ \end{array}\right\} \begin{array}{c} \text{(internal} \\ \text{point P)} \end{array}$$

$$F_{sh} = -\frac{qq_{sh}}{\varepsilon r^2} \qquad\qquad\qquad 0$$

Here ε is a constant which depends on the units employed, and it is assumed that the uniformity of the charge distribution on the shell is not affected by the charge at P, i.e. polarization effects are excluded. If the charges q and q_{sh} are of the same sign the force becomes repulsive and the - sign is replaced by the + sign.

The *gravitational potential* $V_{sh} = U_{sh}/m$ and the *gravitational field* $I_{sh} \equiv F_{sh}/m$ are readily seen to be

$$V_{sh} = -Gm_{sh}/r \text{ (external point) or } -Gm_{sh}/a \text{ (internal point)} \tag{3.13}$$

and

$$I_{sh} = -Gm_{sh}/r^2 \text{ (external point) or } 0 \text{ (internal point)} \tag{3.14}$$

Similarly one can see that in the electrostatic case the potential and the field are

$$V_{sh} = -q_{sh}/\varepsilon r \text{ or } -q_{sh}/\varepsilon a$$

$$I_{sh} = -q_{sh}/\varepsilon r^2 \text{ or } 0.$$

The field is zero inside a uniformly charged and a uniform gravitating spherical shell.

3.5. A SET OF CONCENTRIC SPHERICAL SHELLS

Distinguishing the various shells by a suffix i, so that their radii are $a_1, a_2, \ldots, a_i, \ldots$, we have for an *external* point P that the gravitational potential energy of, and the gravitational force on, a mass m at a distance r from the centre are

$$U = -\frac{Gm}{r} \sum_{\text{all } i} m_{\text{sh},i} = -\frac{GmM}{r}$$

$$F = -\frac{Gm}{r^2} \sum_{\text{all } i} m_{\text{sh},i} \equiv -\frac{GmM}{r^2}$$

where M is the total mass of all the shells.

For an *internal* point, only the shells within the radius r contribute to the force:

$$U_{\text{i}} = -\frac{Gm}{r} \sum_{\substack{i \\ (a_i \leq r)}} m_{\text{sh},i} - Gm \sum_{\substack{i \\ (a_i > r)}} \frac{m_{\text{sh},i}}{a_i}$$

$$F = -\frac{Gm}{r^2} \sum_{\substack{i \\ (a_i \leq r)}} m_{\text{sh},i}.$$

This yields the same result for an external point as if the total mass M were concentrated at the centre and is independent of the radii a_i. For an internal point the contributing mass is

$$\sum_{\substack{i \\ a_i \leq r}} m_{\text{sh},i} \equiv M_r \text{ (say)},$$

and for constant density this is $\left(\frac{4\pi}{3}r^3 / \frac{4\pi}{3}R^3\right)M$. The result (3.2) has therefore been established.

3.6 GAUSS'S THEOREM

Consider now all the points on a fictitious spherical surface S drawn concentrically among the spherical shells. The accelerations are the same in magnitude for all points of S; the potential energies are also the same; so are the gravita-

tional accelerations.   This is so because all these quantities
depend only on the radius of the surface S.   One could have
guessed *this* result without any mathematics, for it follows
from the fact that the physical arrangement of uniform spherical shells gives us no way of distinguishing a preferred
direction: the problem is *spherically symmetrical*.   For the
magnitudes of these quantities one does need some mathematics.

     To formulate these results in a general way, one needs
the outward flux Φ of gravitational acceleration over a closed
surface S.   A typical increment of this, $d\Phi$, is the normal
outward acceleration a multiplied by the incremental area
through which it acts: $d\Phi = a \, dS$ (see Fig. 3.5).   Then one

FIG. 3.5.   An element dS of a closed surface, and the normal outward
acceleration a.

form of Gauss's theorem is (we do not prove it here)

$$\int_{\substack{\text{closed} \\ \text{surface } S}} d\Phi = -4\pi G \text{ (total mass inclosed by } S\text{).} \qquad (3.15)$$

Note that dimensionally the left-hand side is $\left[LT^{-2} . L^2 \right] =$
$\left[L^3 T^{-2} \right]$.   The right-hand side is the same: $[G] [M] =$
$\left[M^{-1} L^3 T^{-2} \right] [M]$.   Applying the theorem to a spherically symmetrical system, the acceleration has the same value (a_r say)
at all points on a spherical surface of radius r with its
centre at the centre of symmetry.   Thus Gauss's theorem yields
in this case

$$\int d\Phi = a_r \int dS = 4\pi r^2 a_r$$

for the left-hand side, and for the right-hand side $-4\pi GM$ or $-4\pi GM_r$. Thus

$$a_r = -\frac{GM}{r^2} \text{ or } -\frac{GM_r}{r^2},\qquad (3.16)$$

in agreement with the results already derived.

3.7. THE GRAVITATIONAL INTERACTION OF TWO SPHERICALLY SYMMETRICAL OBJECTS

The force $-GMm/r^2$ is correct for two interacting particles of masses M and m, a distance r apart. For extended spherically symmetrical masses this result is still valid, since, as far as interaction with an external body is concerned, the mass of each body can be supposed to be concentrated at its centre. To see this, proceed as follows: as far as one body is concerned, reduce the other to a point using eqn (3.2). Then do the same for the remaining body, as shown in Fig. 3.6.

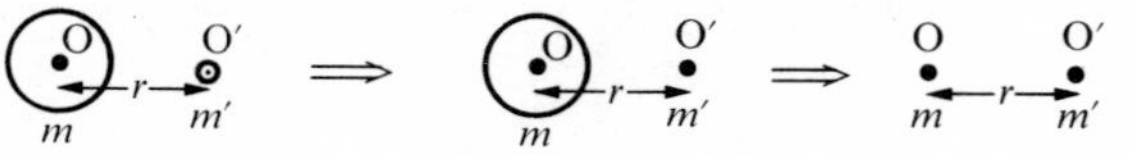

FIG. 3.6. The reduction of the gravitational interaction between two spherically symmetrical masses, centered at 0 and 0', to the interaction between two particles.

Taking positive senses indicated in the figure, the equation for the force $F = -GMm/r^2$ can be used to yield

$$\text{force on } M : M\ddot{r}_M = \frac{GmM}{r^2}, \text{ force on } m : m\ddot{r}_m = -\frac{GmM}{r^2}.\qquad (3.17)$$

This assumes that any motion of the spheres is along the lines of centres, otherwise a centripetal acceleration will feature additionally in eqn (3.17). The accelerations are Gm/r^2, GM/r^2 respectively. The acceleration of m relative to M (see Fig. 3.7) is

$$\ddot{r} = \ddot{r}_m - \ddot{r}_M = -\frac{G}{r^2}(M+m).$$

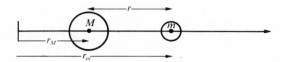

FIG. 3.7. The gravitational interaction between two spherically symmetrical bodies.

This may be written

$$\ddot{r} = -\mu/r^2, \quad \mu \equiv GM(1+m/M) \equiv k^2(1+m/M) \tag{3.18}$$

where μ is sometimes called the 'inertial' constant and k the 'Gaussian' constant. This equation can be integrated by writing it first as

$$2\dot{r}\ddot{r} = -\frac{2\mu}{r^2}\dot{r}, \text{ whence } \frac{\dot{r}^2}{2} - \frac{\mu}{r} = \text{ a constant for the motion.}$$

This looks like an energy equation after multiplying by $m/(1+m/M) \equiv m^*$

$$\tfrac{1}{2}m^*\dot{r}^2 - \frac{GMm}{r} = E \tag{3.19}$$

where E is a constant energy, and the reduced mass m^* incorporates the effect of the motion of M in the kinetic energy term. For large M, the motion of M is negligible and $m^* = m$.

The kinetic energy of the system can be made arbitrarily large by imagining the system of masses M and m to be viewed by an observer in fast uniform motion. Thus, while E is clearly a total energy (kinetic and potential), the question arises: 'For what observer motion is E the correct energy?' It holds for an observer at rest relative to the centre of mass of the system.

The potential energy of the system is

$$U(r) = -GMm/r.$$

At height H above the surface of a spherically

symmetrical body of radius R and mass M the gravitational
acceleration is given by

$$g(r) = -\frac{GM}{r^2} - -\frac{GM}{R^2} \times \left(\frac{R}{r}\right)^2 = g(R)\left(\frac{R}{r}\right)^2$$

where $r \equiv R + H$. If these relations are applied to Sun,
Earth, and Moon, one can plot the accelerations due to these
objects at various distances from them, expressed in units of
$g(R)$, the gravitational acceleration at the surface of the
Earth. The results are shown in Fig. 3.8. The point N marks

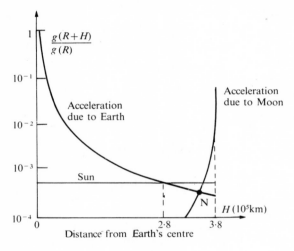

FIG. 3.8. The acceleration caused by the Sun, the Earth, and the Moon at
various distances from the earth, whose radius is denoted by R [1].

a 'neutral' point at which the attraction due to the Earth is
equal to that due to the Moon, but the attraction due to the
Sun at that point is still about twice as large as either of
them separately. The detailed discussion of the attraction
of more than two bodies in mutual gravitational interaction
is very complicated and simplifies only in special cases.
Even the three-body problem touched on here is beyond the
present scope.

3.8. A MATHEMATICAL TECHNIQUE: ENERGY EQUATION OR EQUATION OF MOTION?

3.8.1. *From the energy equation to the equation of motion*

The energy E of a particle of mass m in the field of a spherically symmetrical body of mass M can be written in terms of the distance r between centres. Assuming the particle is outside the body of mass M, and taking a frame of reference in which M is at rest,

$$E = T(r) + U(r) = \tfrac{1}{2}m\dot{r}^2 - \frac{GMm}{r} \tag{3.20}$$

is a constant independent of t. Differentiating with respect to time,

$$m\dot{r}\ddot{r} + \frac{GMm}{r^2}\dot{r} = 0,$$

so that

$$\ddot{r} = -GM/r^2$$

in agreement with eqns (3.1) or (3.16).

3.8.2. *From the equation of motion to the energy equation*

We now start with

$$\ddot{r} = -GM/r^2$$

so that

$$2\dot{r}\ddot{r} = -\frac{2GM}{r^2}\dot{r}.$$

Integrating from $r = r_0$ to r with $(\dot{r})_{r_0} = v_0$ say

$$\dot{r}^2\Big|_{r_0}^{r} = \frac{2GM}{r}\Big|_{r_0}^{r}, \text{ i.e. } \dot{r}^2 - v_0^2 = 2GM\left(\frac{1}{r} - \frac{1}{r_0}\right).$$

Multiplying by $m/2$, we find

$$\tfrac{1}{2}m\dot{r}^2 - \frac{GMm}{r} = \tfrac{1}{2}mv_0^2 - \frac{GMm}{r_0}$$

and this is a constant, independent of time. This is eqn
(3.20) again.

The two elementary but important procedures outlined
above are very useful throughout mathematical physics.

3.9. THE ESCAPE VELOCITY

3.9.1. *Escape assumed*

The energy eqn (3.20) is

$$E = \tfrac{1}{2}mv^2(r) - \frac{GMm}{r} = \tfrac{1}{2}mv^2(\infty)$$

where it has been assumed that the particle can escape to
infinity with velocity $v(\infty)$, as illustrated in Fig. 3.9. This
is possible only if $v^2(\infty) \geqslant 0$. This in turn implies
$v^2(r) \geqslant 2GM/r$, i.e. the velocity at position r from the centre
of the body of mass M satisfies

$$v(r) \geqslant \surd(2GM/r) \equiv v_e(r). \tag{3.21}$$

This equation defines an escape velocity $v_e(r)$ which is seen
to depend on position r. For the Earth at its surface

$$v_e = \left[\frac{2 \times (6.67 \times 10^{-11})\,(m^3kg^{-1}s^{-2})\,(5.98 \times 10^{24})\,(kg)}{6.37 \times 10^6\,(m)}\right]^{\tfrac{1}{2}} = 1.12 \times 10^4 (m\ s^{-1}).$$

Taking the velocity of sound in dry air at normal temperature
and pressure as 332 ms^{-1}, v_e is 33.7 times this velocity, i.e.

$$v_e \sim 33.7\ \text{Mach},$$

which is much larger than the speed of our fastest aeroplanes.

3.9.2. *No escape possible*

If $v(r) < v_e(r)$, the particle can reach only a distance
r_{max} from the attracting body which is given by the vanishing
of the kinetic energy, as shown in Fig. 3.9. Thus if there
is escape we can talk about a velocity $v(\infty)$ which is zero or
positive. If no escape is possible, one can talk about a
maximum distance r_{max} from the centre.

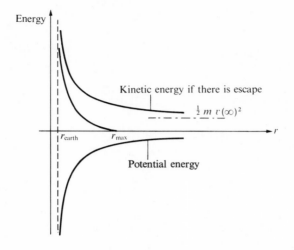

FIG. 3.9. The kinetic energies of two particles projected from the Earth's surface. The top curve refers to a particle which can escape; the curve below it refers to a particle which falls back to the Earth.

Having defined the escape velocity, we can express $v(\infty)$ and r_{max} rather simply as follows:

$$v^2(\infty) = v^2(r) - v_e^2(r),$$

$$r_{max} = \frac{r v_e^2(r)}{v_e^2(r) - v^2(r)}$$

(3.22)

The expression for $v^2(\infty)$ derives directly from energy conservation and is therefore valid even when the motion is not in a straight line, but the expression for r_{max} assumes straight-line motion.

3.10. THE CLASSICAL BLACK HOLE

Imagine a spherically symmetrical set of gravitating particles which come together under their interactions. If the total mass M is fixed, then as the system shrinks the escape velocity increases in accordance with eqn (3.21). At a limiting radius $r = r_0$ the escape velocity $v_e(r_0)$ reaches the speed of light. At this stage no particle can escape from the system, since one cannot project it with a greater velocity. This also applies to the photon (a 'particle' of light energy) even though it has no rest mass. It can be thought of interacting

TABLE 3.1

Solar system data

Body	Distance from the Sun, $r(10^{12}\text{m})$	Radius of body, $r_0(10^8\text{m})$	Density (10^3kg m^{-3})	Escape velocity $v_e(\text{m s}^{-1})$	Gravitational acceleration, $g_0(\text{m s}^{-2})$	Orbital eccentricity, ε
Sun	–	6.955	1.41	$6.18.10^5$	2.73	–
Mercury	0.05795	0.02439	5.4	$4.25.10^3$	3.70	0.2056
Venus	0.1082	0.06050	5.21	$1.04.10^4$	8.86	0.0068
Earth	0.1497	0.06378	5.52	$1.12.10^4$	9.81	0.0167
Moon	3.844×10^{-4}*	0.01738	3.34	$2.37.10^3$	1.62	0.0549
Mars	0.2281	0.03400	3.94	$5.03.10^3$	3.72	0.0933
Jupiter	0.7789	0.7135	1.34	$5.96.10^4$	25.90	0.0484
Saturn	1.429	0.6040	0.69	$3.67.10^4$	11.10	0.0557
Uranus	2.868	0.236	1.60	$2.22.10^4$	10.45	0.0472
Neptune	4.493	0.246	1.60	$2.35.10^4$	11.20	0.0086

*Distance from the Earth.

with a gravitational field with a mass E/c^2 where E is its energy. Thus the condition $v_e(r_0) = c$ implies that not even light can leave the system, which thus becomes invisible, explaining why it is then called a 'black hole'. The required radius is

$$r_0 = 2GM/c^2. \tag{3.23}$$

In general relativity this expression is found to have a similar significance and is called the radius of a Schwarzschild black hole. For the Sun ($M = 2 \times 10^{30}$ kg), $r_0 \sim 3$ km.

A black hole can still be sensed by its gravitational effects, and gas clouds and other objects could fall into it. In this process such clouds would heat up and radiate energy, which can be detected at a distance if it is emitted outside the surface $r = r_0$.

The theoretical investigation of black holes is currently fashionable, but none have so far been identified experimentally with certainty.

3.11. SOLAR SYSTEM DATA
For interest, some data about the solar system is given in Table 3.1.

PROBLEMS

(3.1)   Spherically symmetrical bodies of masses m_1, m_2 have centre coordinates r_1, r_2 at a given instant. The corresponding centre-of-mass coordinate R, measured along the line of centres, is

$$(m_1 + m_2)R = m_1 r_1 + m_2 r_2.$$

The total kinetic energy of the system is defined to be $\frac{1}{2}(m_1 \dot{r}_1^2 + m_2 \dot{r}_2^2)$. If $r = r_1 - r_2$, show that the kinetic energy is

$$\tfrac{1}{2}(m_1+m_2)\dot{R}^2 + \tfrac{1}{2}\frac{m_1}{1+m_1/m_2}\,\dot{r}^2$$

Interpret this result.

(3.2) The virial theorem states that if for a system of finite extent the potential between any two particles at a
distance r apart varies as r^s, where s is a constant,
then the (time) average kinetic energy $\bar{T}$ and the average
potential energy $\bar{U}$ are related by $2\bar{T} = s\bar{U}$. Let $\bar{E}$ be the
average total energy.

   (i)   Show that $\bar{U} = (2/s+2)\bar{E}$, $\bar{T} = (s/s+2)\bar{E}$

   (ii) For a mass m in a circular orbit of radius a about
         a stationary central mass M show that the accelera-
         tion towards the centre is $g(a) = GM/a^2 = v^2(a)/a$.
         Hence show that

$$\bar{T} = GMm/2a,\ \bar{U} = GMm/a,\ \bar{E} = GMm/2a.$$

         Does the virial theorem hold, and what is the
         value of s?

(3.3) A spherically symmetric planet has mean density $\bar{\rho}$,
radius R and mass $M = \frac{4}{3}\pi\bar{\rho}R^3$. A satellite describes a
circular orbit of radius r about the planet, with orbi-
tal speed $v_c(r)$ and orbital period $T(r)$. The gravitational acceleration at the surface of the planet is g.

   (i)   Show that for an orbit skimming the surface of the
         planet ($r = R$)

$$v_c(R) = \surd(gR),\ T(R) = 2\pi\surd(R/g)$$

         and that $g = GM/R^2$. Deduce that $T(R)$ is the same
         for planets of different sizes if their mean den-
         sity is the same. Evaluate $v_c(R)$ and $T(R)$ for the
         Earth, using the approximate values $g \sim 10$ m s^{-2}
         and $R \sim 6.4 \times 10^6$ m.

   (ii) Show that in general

$$v_c(r) = \sqrt{(gR^2/r)} \text{ and } v_e(r) = \sqrt{2}\, v_c(r)$$

where $v_e(r)$ is the escape velocity defined in eqn
(3.21). Estimate the orbital speed and period for
a satellite such as Skylab (1973) given that its
height above the Earth's surface is about 400 km.

(iii)   Show also that

$$T(r) = T(R)(r/R)^{3/2}.$$

Is this Kepler's third law?

(iv)   Communication and television relay satellites such
as Early Bird (1965) are often launched into an
orbit which is synchronous, i.e. the satellite's
orbital period $T(r)$ equals the Earth's period of
rotation which is about 23 hours 56 minutes (why
not 24 hours?). Show that only one value of r
gives a circular synchronous orbit, and find this
value.

(v)   Mars has two satellites, Phobos and Deimos, in
near-circular orbits with radii of 9300 km and
23 400 km respectively. The orbital period of
Phobos is about 7 hours 42 minutes. Deduce the
period of Deimos and the surface gravity of Mars
(the radius of Mars is 3400 km), and compare the
latter result with that in Table 3.1.

(vi)   The Earth's orbital speed around the Sun is about
30 km s^{-1}. By (ii) of this question a space probe
leaving the vicinity of the Earth with a speed of
30$\sqrt{2}$ km s^{-1} *relative to the Sun* will escape the
Sun's gravitational field and leave the solar
system. If the launch direction is chosen care-
fully this need only mean leaving the vicinity of
the Earth with a speed *relative to the Earth* of
$v(\infty) = (\sqrt{2} - 1)30$ km s^{-1}. Using eqn (3.22) and
the numerical results of part (i) show that the
space probe must have a speed on leaving the
Earth's atmosphere ($r \sim R$) which is about 50 per
cent higher than the speed needed simply to escape

from the Earth.

(3.4) Taking GM for the sun to be $1.329 \times 10^{20} \text{m}^3 \text{s}^{-2}$ and the Earth's distance from the Sun in its slightly elliptic orbit as $1.498 \times 10^{11} \text{m}$, calculate the periodic time of revolution of the Earth around the Sun, treating the motion as circular.

(3.5) A thick spherical shell of gravitating matter has internal and external radii a and b, and its density varies inversely as the distance from the centre: $\rho(r) = k/r$, where k is a constant. The shell is imagined as built up by the condensation from infinity of successive thin spherical shells on the layers already present.

   (i)   When built up to a radius r, show that the amount of matter already condensed is

$$M(r) = 2\pi k(r^2 - a^2).$$

   (ii) When built up to a radius r show that the potential at an external point at distance r is

$$V(r) = -2\pi k G(r^2 - a^2)/r.$$

   (iii) Explain why the work done by the field in condensing a shell of mass $dM(r)$ and of thickness dr on a thick shell extending from radius a to radius r is

$$dW(r) = -V(r)dM(r) = -dU(r)$$

   where $dU(r)$ is the increase in potential energy.

   (iv) Evaluate the total work done by the field in condensing the thick shell,

$$\int_{r=a}^{r=b} dW(r) = \frac{2G}{3} M^2 \frac{b+2a}{(b+a)^2}$$

   where M is the total mass of the shell, and explain the sign.

(v)   What is the total gravitational potential energy of the shell? Explain the sign.

(3.6)   A spherically symmetrical non-uniform gravitating sphere has a mean mass density $\rho(r)$ up to a radius r and a surface mass-density $\sigma(r)$ at radius r, where $\rho(r)$ and $\sigma(r)$ are appropriate functions of r.

(i)   Show by considering the addition of a thin layer of density $\rho(r)$ on the sphere already built up to a radius r, that

$$\frac{1}{3}\left(\rho + r\frac{d\rho}{dr}\right) = \sigma - \frac{2}{3}\rho$$

(ii)   If $g(r)$ is the gravitational acceleration at radius r, show that

$$-g(r) = \frac{4\pi}{3}rG\rho(r)$$

and

$$-\frac{dg(r)}{dr} = 4\pi G\left(\sigma - \frac{2}{3}\rho\right).$$

(iii)   The mean density of the Earth is $\rho(R) \sim 5520$ kg m^{-3}. The density of the Earth near the surface is, apart from the heterogeneous crust, 3300 kg m^{-3}. Explain how one would expect the *magnitude* of the gravitational acceleration to change as one goes down a deep shaft in surroundings which are similar to the *average* surface conditions of the earth. Sketch the curve of $g(r)$ against r which you would expect.

(3.7)   Taking D to be the mean distance between Earth and Moon, obtain an approximate equation for the distance x from the Moon at which the neutral point occurs. The approximations to be made are that both bodies are to be treated as uniform spheres, and the attraction due to the Sun is to be treated as a constant over the whole of the

distance D.   Show that, if E, M are the masses of the Earth
and the Moon,

$$\frac{x}{D} = \frac{\sqrt{(E/M)} - 1}{(E/M) - 1}$$

If $E = 5.95 \times 10^{24}$ kg, $M = 7.35 \times 10^{22}$ kg, $D = 3.8 \times 10^5$ km,
find x.

4

THE COSMOLOGICAL DIFFERENTIAL EQUATION:
THE PARTICLE MODEL

4.1. THE QUALITATIVE PICTURE

Most people are familiar with a microscope. If one uses it to look at a simple object such as a blade of grass, it reveals details not visible to the naked eye. As the power of the microscope is increased, a network of veins and cells comes into view revealing the intricacies and complication of what was first considered to be a simple object.

We are less familiar with the reverse operation of taking a rougher instead of a finer view of our surroundings. If we photograph a growing plant once a day and compress the time scale by playing the resulting film back, we can see in a few seconds how the plant grew in a year. Instead of compressing in time one can also compress in space: an aerial view of large fields may reveal regular features not obvious to a person on the ground. For example, a certain unevenness on the ground may have a geometric form, and this may lead one to discover that below the field lies a prehistoric city.

In an important branch of cosmology one is concerned with such compressions in space and time. By ignoring the complications of solar systems and galaxies and taking a very rudimentary and rough picture of the universe, one is trying to trace its development in space and time. This type of treatment represents a compression because it seeks by simple equations to reveal the large-scale properties of the universe, while ignoring all finer details. The discussion of these details can then be introduced at a later stage to yield a more complete or better theory. In this way it is possible to discuss the formation of simple atoms, of galaxies and so on, as indicated in Section 1.4. These considerations require difficult calculations, and are part of current research in astrophysics and cosmology.

4.2. THE ENERGY TERMS

We shall assume a finite model universe and shall discuss
infinite universes later (Section 5.3). This model can be
pictured as a very large assembly of clumps of matter in
motion. These clumps of matter may be taken to be galaxies.
It is important that they are electrically neutral, so that
they do not interact electrically; but they do interact
gravitationally. Suppose three such clumps of matter of
masses m_1, m_2, m_3 to be situated at a time t at positions
$r_1(t), r_2(t), r_3(t)$, as measured from a fixed origin O. If
their velocities are radial and of magnitude $\dot{r}_j(t)$ ($j = 1,2,3$),
the kinetic energy of the system is

$$T = \frac{1}{2}(m_1 \dot{r}_1^2 + m_2 \dot{r}_2^2 + m_3 \dot{r}_3^2) = \frac{1}{2} \sum_{j=1}^{3} m_j \dot{r}_j^2.$$

The gravitational potential energy of any pair of masses is
negative and given by

$$-G \frac{m_1 m_2}{|\mathbf{r}_1 - \mathbf{r}_2|}, \quad -G \frac{m_1 m_3}{|\mathbf{r}_1 - \mathbf{r}_3|}, \quad -G \frac{m_2 m_3}{|\mathbf{r}_2 - \mathbf{r}_3|},$$

where G is Newton's gravitational constant. The total potential energy is therefore

$$V_G = -G \left[\frac{m_1 m_2}{r_{12}} + \frac{m_1 m_3}{r_{13}} + \frac{m_2 m_3}{r_{23}} \right] = -G \sum_{\substack{i<j \\ i,j=1}}^{3} \frac{m_i m_j}{r_{ij}}$$

where $r_{ij} \equiv |\mathbf{r}_i - \mathbf{r}_j|$. The constraint $i < j$ indicates that one
does not wish to have the two terms $m_i m_j / r_{ij}$ and $m_j m_i / r_{ji}$ in
the sum, but only one of them (for example the term for which
i is less than j). The total energy of the system is accordingly

$$E = T + V_G = \frac{1}{2} \sum_{i=1}^{n} m_i \dot{r}_i^2 - G \sum_{\substack{i<j \\ i,j=1}}^{n} \frac{m_i m_j}{r_{ij}}, \qquad (4.1)$$

where we have generalized to a system of n clumps of matter.
Note that the potential energy becomes less negative, i.e. it
increases as the particle separations go up (see Fig. 4.1).

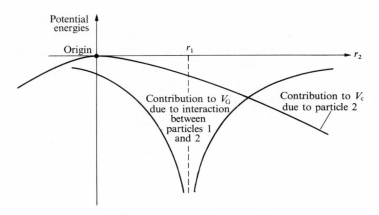

FIG. 4.1. Potential energies due to the cosmological force acting on particle 2, and due to the gravitational interaction of particle 2 with the fixed particle 1, as a function of position of particle 2.

For some purposes it is desirable to add a further potential energy term $-(\lambda/6)m_j r_j^2$ for each particle. The factor 1/6 has been inserted for convenience. This yields an additional potential energy

$$V_c = -\frac{\lambda}{6} \sum_j m_j r_j^2 \tag{4.2}$$

arising from a force

$$F_{cj} = \frac{\lambda}{3} m_j r_j \tag{4.2'}$$

on particle j, where λ is a constant. Let us first take it to be positive; then eqn (4.2') is positive, and the force is repulsive. This may also be seen from eqn (4.2): as the particle separations increase, V_c becomes more negative, i.e. it *decreases*. The behaviour of V_c is therefore opposite to that of V_G. Whereas V_G encourages the particles to be close together so that the system can approach a state of low potential energy, V_c reaches a low value if the particles are far apart. Thus one sees again that the gravitational interaction is attractive, whereas the new interaction V_c acts in the opposite sense and seeks to remove the particles to infinity.

But since it does not depend on *two* physical parameters, such
as *two* interacting masses or *two* interacting charges, it is
not a 'repulsion' in the normal sense. One can, however, call
it a *cosmic repulsion*. This repulsion is still rather con-
troversial. We shall be able to discuss it since it introduces no undue complications and it frequently occurs in
modern work. The reader may, however, eliminate it from all
subsequent equations without difficulty by putting λ equal to
zero. He may alternatively interpret it as a *cosmic attraction* by regarding λ as a negative constant. This is also
sometimes done in current work on cosmology. The choice of
the power 1 in eqn (4.2') and hence of the power 2 in eqn
(4.2) will be discussed in Section 5.3.

   The reason why the cosmic repulsion was first introduced
is as follows. Prior to the 1920s which brought Hubble's law,
the universe was believed to be in a broadly static and equilibrium condition by many scientists. Now a theory based on
movement (kinetic energy) and attraction (gravitational
forces) cannot yield a static model. Either the kinetic
energy dominates and the universe is for ever expanding, or
the energy of motion of the galaxies is inadequate to overcome
the mutual gravitational attraction, in which case the system
will collapse upon itself. Neither possibility is a static
one. In fact the two possibilities correspond physically to
either projecting a rocket beyond the Earth's attraction, or
throwing a ball into the air which then returns to Earth.
These Newtonian cosmological ideas were developed only in the
1930s. They have, however, their counterpart in general relativity, and were first noted in that context by Einstein when
he created the theory in 1915. In order to balance the gravitational attraction and produce a static model he needed a
repulsive force, and so he introduced the cosmological repulsion. He withdrew it again in 1931, when the expansion of the
universe became an accepted proposition. Einstein, had he
believed the general theory of relativity quite uncompromis-
ingly and not introduced the cosmological force, would have
predicted an expanding or a contracting universe. The cosmological constant has been used since that time, and has had

alternating periods of popularity. We shall keep it in our
equations, particularly as there is currently some evidence
for a non-zero value [1,2].

4.3. THE EFFECT OF HUBBLE'S LAW

In order to incorporate Hubble's law into our scheme, it will
be assumed that at any time t the galaxies have a radial
motion which is given by

$$r_i(t) = \bar{R}(t) r_i(t_0) \quad (i = 1,2,\ldots,n) \tag{4.3}$$

where $\bar{R}(t)$ is a universal function of time, *the same for all
the particles*, defined in eqn (2.3). The time t_0 is some
standard time to which all observations may be referred and
is usually the present time. The (radial) velocity of the
ith particle is

$$\frac{\mathrm{d}r_i(t)}{\mathrm{d}t} = \frac{\mathrm{d}\bar{R}(t)}{\mathrm{d}t} r_i(t_0) = \frac{\mathrm{d}\bar{R}(t)/\mathrm{d}t}{\bar{R}(t)} r_i(t). \tag{4.4}$$

Denoting time differentiation by a dot, and Hubble's parameter
by $H(t)$,

$$\dot{r}_i(t) = H(t) r_i(t), \quad H(t) \equiv \dot{\bar{R}}(t)/\bar{R}(t). \tag{4.5}$$

Thus Hubble's parameter determines the same connection between
velocity and position for all particles i *at the same time* t.
It is not a constant in the sense that this connection will in
general vary in time: H, like $\bar{R}$, is independent of the suffix
i, but it does depend on t. Thus eqn (4.3) effects a factori-
zation of a function of i and t into a product of two functions, one depending on i and the other depending on t.

4.4. THE ENERGY EQUATION FOR n PARTICLES

The energy of the n particles is, by eqns (4.1) and (4.2),

$$E = \frac{1}{2} \Sigma m_i \dot{r}_i^2 - G\Sigma' \frac{m_i m_j}{r_{ij}} - \frac{\lambda}{6} \Sigma m_i r_i^2$$

where Σ denotes $\displaystyle\sum_{i=1}^{n}$ and Σ' denotes $\displaystyle\sum_{\substack{i,j=1 \\ (i<j)}}^{n}$. Incorporating eqns

(4.3) and (4.4), one finds

$$T = \frac{1}{2} \Sigma m_i \left[\dot{\bar{R}} r_i(t_0) \right]^2 = A\dot{\bar{R}}^2$$

$$V_G = -G\Sigma' \frac{m_i m_j}{\bar{R} r_{ij}(t_0)} = -\frac{B}{\bar{R}}$$

$$V_c = -\frac{\lambda}{6} \Sigma m_i \left[\bar{R} r_i(t_0) \right]^2 = -D[\bar{R}]^2$$

so that

$$E = A\dot{\bar{R}}^2 - \frac{B}{\bar{R}} - D\bar{R}^2. \tag{4.6}$$

Since the energy E of the n-particle system is a constant, eqn (4.6) represents a constraint on the possible time-variation of R. The coefficients are

$$A \equiv \frac{1}{2} \Sigma m_i \left[r_i(t_0) \right]^2 \quad \text{(energy} \times \text{(time)}^2) \tag{4.7}$$

$$B \equiv G\Sigma' \frac{m_i m_j}{r_{ij}(t_0)} \quad \text{(an energy)} \tag{4.8}$$

$$D \equiv \frac{\lambda}{6} \Sigma m_i [r_i(t_0)]^2 \quad \text{(an energy).} \tag{4.9}$$

They are independent of time, and depend only on the nature of the system at the reference time t_0. In eqn (4.6) we have derived the important *cosmological differential equation* for the scale factor $\bar{R}(t)$.

Eqn (4.6) has a simple interpretation. Neglecting the term due to cosmic repulsion, an expanding system (e.g. a universe) increases its potential energy $-B/\bar{R}(t)$, for as $\bar{R}$ increases, this term becomes less negative. Thus the total energy remains constant only if the kinetic energy $A(\dot{\bar{R}})^2$ also decreases with expansion. This is reasonable since the constituent clumps of matter have to overcome gravitational

attractions in order to participate in the expansion. They
are therefore slowed down. How much are they slowed down?
How long will the expansion last? These are typical questions
for cosmology. Some rough answers can be given, and this will
be done below.

4.5. THE RESCALED ENERGY EQUATION

If we note from eqn (4.7) and (4.9) that

$$D/A = \lambda/3, \tag{4.10}$$

the energy eqn (4.6) can be written more elegantly as

$$\dot{\bar{R}}^2 = \frac{B}{A\bar{R}} + \frac{\lambda}{3} \bar{R}^2 + \frac{E}{A} \tag{4.11}$$

Thus, even if G and λ are universal constants, one still
obtains a different differential equation for each value of
B/A and E/A. By passing to a new function R, instead of $\bar{R}$,
it proves possible to collect the main physical characteristics of the model into a new constant C, and, with reference
to the energy E, to retain merely the sign of it. This is
quite adequate since the property of chief interest is the
nature of the evolution of the universe in time. This is
clearly unaffected if P is independent of time in

$$R(t) = P\bar{R}(t), \tag{4.12a}$$

which will be used to rescale the cosmological differential
equation. This relation implies

$$r_i(t) = \frac{r_i(t_0)}{P} R(t) \equiv s_i R(t) \tag{4.12b}$$

where s_i is a coordinate which is constant in time for each
galaxy and will be used later (Chapter 11).

We shall now pursue the consequences of (4.12). Using
(4.12a) in (4.11),

$$\frac{\dot{R}^2}{P^2} = \frac{BP}{A} \frac{1}{R} + \frac{\lambda}{3P^2} R^2 + \frac{E}{A}$$

i.e.,

$$\dot{R}^2 = \frac{BP^3}{A} \frac{1}{R} + \frac{\lambda}{3} R^2 + \frac{EP^2}{A}. \tag{4.13}$$

Let P now be chosen so as to obtain agreement with general relativistic equations as normally presented:

(a) If $E>0$, choose $P^2 = c^2 A/E$, i.e. in $EP^2/A = -kc^2$ choose $k = -1$.

(b) If $E<0$, choose $|P^2 = c^2 A/|E|$, i.e. in $EP^2/A = -kc^2$ choose $k = +1$.

(c) If $E = 0$, leave P arbitrary, i.e. in $EP^2/A = -kc^2$ choose $k = 0$.

Thus
$$P^2 = \frac{c^2 A}{|E|} \ (E \neq 0).$$

$$\left. \right\} (4.14)$$

The cosmological differential eqn (4.13) is then

$$\dot{R}^2 = \frac{C}{R} + \frac{\lambda}{3} R^2 - kc^2 \ (k = -1, \ 0 \ \text{or} \ +1) \tag{4.15}$$

$$C \equiv \frac{BP^3}{A} = \frac{c^2 BP}{|E|} \ (E \neq 0). \tag{4.16}$$

If c is the velocity of light *in vacuo*, eqn (4.15) is precisely a general relativistic result for the simplest cosmological models. Its derivation here as a classical result represents a development of later ideas [3,4,5]. Note that we are still dealing with individual galaxies as shown by eqns (4.7) to (4.9).

Applying dimensional analysis, and denoting the dimensions of A by $[A]$,

$$[P^2] = \left[\frac{c^2 A}{E}\right] = [L^2 T^{-2}] \frac{[ML^2]}{[ML^2 T^{-2}]} = [L^2].$$

It follows that, while $\bar{R}$ is dimensionless, R is a length. Eqn (4.15) is the cosmological differential equation which contains as main parameters C, $k = -1$ or 0 or $+1$, and λ. Its properties will occupy us a great deal in the following chapters.

One should note that by eqn (4.3),

$$\bar{R}(t_0) \equiv \bar{R}_0 = 1. \qquad (4.17)$$

Thus eqn (4.12) furnishes

$$R(t_0) \equiv R_0 = P = (-kc^2 A/E)^{\frac{1}{2}} = c(A/|E|)^{\frac{1}{2}} \qquad (4.18)$$

where the last two forms hold if $E \neq 0$. Also

$$H \equiv \frac{\dot{\bar{R}}}{\bar{R}} = \frac{\dot{R}}{R}. \qquad (4.19)$$

Therefore, if one uses eqn (4.15) at the characteristic time t_0, one finds, after dividing by P^2 and using (4.14),

$$H_0^2 = \frac{B}{A} + \frac{\lambda}{3} + \frac{E}{A}. \qquad (4.20)$$

The last term can also be written as $-kc^2/P^2$. This is a very useful result and becomes more so, if one makes the identification that

$$t_0 \text{ is the present time,} \qquad (4.21)$$

since the left-hand side of eqn (4.20) is then reasonably well known from experiment. We shall adopt the interpretation (4.21) throughout the rest of this book.

4.6. PARAMETERS FOR THE EVOLUTION OF THE UNIVERSE

A number of parameters have become standard in the description of the universe. One is the Hubble parameter (4.19). Another variable measures the rate at which the expansion of the universe slows down, and is called the *deceleration parameter*. As such it is made to involve $-\ddot{\bar{R}}$. To make it dimensionless it is then defined by

$$q \equiv -\frac{\bar{R}\ddot{\bar{R}}}{\dot{\bar{R}}^2} = -\frac{R\ddot{R}}{\dot{R}^2} \qquad (4.22)$$

It can be calculated from the 'equation of motion of the

universe' which is understood to be the time derivative of the energy eqn (4.15):

$$\ddot{R} = -\frac{C}{2R^2} + \frac{\lambda}{3} R.$$

(4.23)

The left-hand side corresponds in classical mechanics to the acceleration, while the right-hand side gives the cosmological force (4.2') and the gravitational force which act on unit mass. Using eqn (4.19).

$$q = -\frac{\ddot{R}R}{(RH)^2} = -\frac{1}{RH^2} \left[-\frac{C}{2R^2} + \frac{\lambda}{3} R \right] = \sigma - L$$

(4.24)

We have here introduced the *density parameter* representing the deceleration due to gravity which is, using eqn (4.16),

$$\sigma \equiv \frac{C}{2H^2 R^3} = \frac{B}{2AH^2} \left(\frac{P}{R} \right)^3, \text{ so that } \sigma_0 = \frac{B}{2AH_0^2}.$$

(4.25)

We have also introduced a parameter representing the acceleration of the expansion of the universe due to the cosmological force:

$$L \equiv \frac{\lambda}{3H^2}, \text{ so that } L_0 = \frac{\lambda}{3H_0^2}.$$

(4.26)

The parameters thus defined vary in their accessibility to observation. The present value of H is known with fair accuracy from the slope of the red-shift versus luminosity distance curve (Section 11.2); the present value of the mean density ρ of matter, and hence of σ, can be estimated from the observed density of matter in stars, dust, and gas, though the correct value of ρ may well be higher (Section 6.2.3). It is in principle possible to estimate q from the red-shift versus luminosity distance curve, as discussed in Section 11.5 but the uncertainties are large.

We give below some additional relations which hold also in general relativity and are sometimes useful. A relation between σ and q arises from the energy eqn (4.15) which can be written as

$$\frac{kc^2}{R^2 H^2} = \frac{kc^2}{\dot{R}^2} = \frac{C}{R\dot{R}^2} + \frac{\lambda}{3}\frac{R^2}{\dot{R}^2} - 1 = \frac{C}{H^2 R^3} + \frac{\lambda}{3H^2} - 1.$$

Defining an *energy parameter* K, this may be written as

$$K \equiv \frac{kc^2}{R^2 H^2} = 2\sigma + L - 1 = 3\sigma - q - 1. \qquad (4.27)$$

Thus eqns (4.24) and (4.26) are alternative ways of writing the equation of motion and the energy equation respectively. They determine the evolution of the universe in time.

There are also some purely *kinematic links* between these parameters, i.e. links which do not involve the forces: for example

$$\dot{H} = \frac{\ddot{R}}{R} - \frac{\dot{R}^2}{R^2} = -H^2 q - H^2 = -(1+q)H^2. \qquad (4.28)$$

If $E \neq 0$ one can also obtain expressions for P and C. By eqns (4.13) and (4.20)

$$P = c\left[\frac{|E|}{A}\right]^{-\frac{1}{2}} = c\left|H_0^2 - \frac{B}{A} - \frac{\lambda}{3}\right|^{-\frac{1}{2}} = \frac{c}{H_0}\left|1 - 2\sigma_0 - L_0\right|^{-\frac{1}{2}}. \qquad (4.29)$$

Now using (4.16),

$$C = \frac{B}{A}P^3 = \frac{2\sigma_0 c^3}{H_0}\left|1 - 2\sigma_0 - L_0\right|^{-\frac{3}{2}} \qquad (4.30)$$

It will be shown in the next chapter (eqn (5.8)) that in a continuum approximation C can be expressed in terms of P and a mean density of matter to be denoted by $\rho(t)$.

PROBLEMS

(4.1)   Prove that galaxies cannot overtake each other in the model adopted in this chapter.

(4.2)   Discuss models of constant deceleration parameter q and show that for $q \neq -1$ and $R = 0$ at $t = 0$ there exists a constant A such that

$$R^{q+1} = At, \quad H = \frac{1}{(q+1)t}.$$

For $q = -1$ show that

$$R(t) = R(t_0) \exp H(t-t_0).$$

[The case $q = -1$ corresponds to the steady-state model of Section 2.4; the case $q = 0$, $H = 1/t$ corresponds to the Milne model of Section 2.5; and the case $q = \frac{1}{2}$, $H = 2/3t$ corresponds to the Einstein—de Sitter model of Section 2.6.]

(4.3) [6] Show from eqn (4.15) and its first and second derivatives that

$$\frac{\lambda}{3} = H^2 + \frac{kc^2}{R^2} - \frac{C}{R^3}$$

$$= \frac{C}{2R^3} - qH^2$$

$$= QH^2 - \frac{C}{R^3}$$

where a third-derivative parameter has been introduced:

$$Q \equiv \dddot{R}R^2/\dot{R}^3$$

(4.4) [7] Show from problem (4.3) that

$$\frac{kc^2}{R^2} = \frac{3C}{2R^3} - (q+1)H^2 = (Q-1)H^2$$

and

$$\lambda = (Q-2q)H^2, \quad \frac{3}{2}\frac{C}{R} = (Q+q)H^2.$$

Hence verify that $q < \frac{1}{2}$ for ever-expanding ($\lambda = 0$) models is a condition which corresponds in some sense to $Q < 1$ for models with $\lambda \neq 0$.

5

THE COSMOLOGICAL DIFFERENTIAL EQUATION:
THE CONTINUUM MODEL

5.1. THE CONTENTS OF THE UNIVERSE AS A UNIFORM CONTINUUM

In order to estimate the coefficients, a picture of the gross
features of the universe will be adopted which regards all the
matter in it as smeared out uniformly at any one time. The
matter density $\rho(t)$ (mass per unit volume) becomes the main
function in terms of which the universe can be described. It
is then convenient to assume that the model universe is
bounded by a spherical surface S whose radius at any time t
will be denoted by $a(t)$. The origin of coordinates will be
taken at the centre 0 of this sphere.

At the reference time t_0 a spherical shell of radius x
and thickness dx has surface area $4\pi x^2$ and volume $4\pi x^2\, dx$.
The amount of matter in it is

$$dm(x) = 4\pi x^2 \rho_0\, dx. \tag{5.1}$$

Now consider any function $f(r_i)$ of the n particles within S:

$$\Sigma m_i f(r_i) \rightarrow \int_0^{a_0} 4\pi x^2 \rho_0 f(x)\, dx. \quad [a_0 \equiv a(t_0)] \tag{5.2}$$

We have on the right-hand side the estimate of this sum in the
approximation of a uniformly smeared out universe. Instead
of summing $m_i f(r_i)$ over all the particles, one has here integrated over all the spherical shells from radius $x = 0$ to
$x = a_0$. At $x = 0$ the shell has shrunk to a point which does
in fact not contribute to the integral for most reasonable
functions $f(x)$. Applying this process to (4.7),

$$A = \tfrac{1}{2} \int_0^{a_0} 4\pi x^2 \rho_0 x^2\, dx = \frac{2\pi}{5} \rho_0 a_0^5 = \frac{3M_T}{10} a_0^2. \tag{5.3}$$

The total mass of the model universe has here been introduced:

$$M_T \equiv \Sigma m_i \rightarrow \int_0^{a_0} 4\pi x^2 \rho_0\, dx = \frac{4\pi}{3} \rho_0 a_0^3. \tag{5.4}$$

Since this quantity is constant in time one has $\rho a^3 = \rho_0 a_0^3$, i.e. using eqn (4.3) with $r_i(t) = a(t)$ and also using eqn (4.12),

$$\frac{\rho_0}{\rho} = \left(\frac{a}{a_0}\right)^3 = \bar{R}^3 = \left(\frac{R}{P}\right)^3. \qquad (5.5)$$

The coefficient D of eqn (4.9) could be calculated similarly, but is more easily found from eqns (4.10) and (5.3). We now evaluate B (Fig. 5.1).

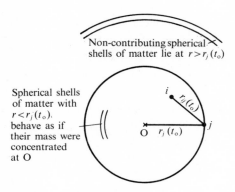

FIG. 5.1. The evaluation of B. In eqn (5.6), $r_j(t_0)$ has been denoted by x, since it is there a variable of integration.

It is now convenient to imagine that the particles are so numbered that $i < j$ if, and only if, particle i is no further from the centre than particle j. This means that the gravitational potential at the location of particle j, at time t_0, due to all particles i lying closer to the centre than j, is

$$V[r_j(t_0)] = -G \sum_{i=1}^{j-1} \frac{m_i}{r_{ij}(t_0)}.$$

With this definition eqn (4.8) can be written

$$B = -\sum_{j=1}^{n} m_j \, V[r_j(t_0)].$$

Because of spherical symmetry with respect to 0 the potential is a function only of distance r_j from 0. In a continuum model eqn (4.8) can then be expressed as an integral over the distance $r_j = x$, as in eqn (5.2):

$$B = - \int_0^{a_0} 4\pi x^2 \rho_0 \, V[x,t_0] \, dx. \qquad (5.6)$$

Note that the time now enters into V independently of x. The reason is that the effect of all the particles $i < j$ has been expressed in terms of the matter density ρ_0 which decreases as the universe expands. V is the gravitational potential at distance x from 0 due to all the mass $M(x)$ inside a sphere of radius x. Therefore, from Section 3.5 (with $V = U/m$)

$$V(x,t_0) = -\frac{GM(x)}{x} = -\frac{4\pi}{3} G\rho_0 x^2.$$

Substituting in eqn (5.6), one finds

$$B = \frac{16\pi^2}{15} G\rho_0^2 a_0^5. \qquad (5.7)$$

Note that B and D are the total gravitational and cosmological potential energies respectively of the universe at the present time t_0. Now substitute eqns (5.3) and (5.6) into (4.16) to find

$$C = \frac{B}{A} P^3 = \frac{8\pi}{3} G\rho_0 P^3 = \frac{8\pi}{3} G\rho R^3, \qquad (5.8)$$

where (5.5) has also been used.

The continuum model therefore yields (4.15) with (5.8), i.e.

$$\dot{R}^2 = \frac{C}{R} + \frac{\lambda}{3} R^2 - kc^2, \quad C \equiv \frac{8\pi}{3} G\rho R^3. \qquad (5.9)$$

These are the precise results obtained for the simplest models of the universe from relativistic cosmology. *This identification of the constant C is the main contribution made to our model by passing from the particle picture to the approximation of the content of the universe by a continuous matter distribution.*

5.2. THE COSMOLOGICAL DIFFERENTIAL EQUATION FROM THE EQUATION OF MOTION

We now give an alternative derivation of the cosmological differential equation. By Newton's laws, a spherically symmetric distribution of mass with centre 0 acts on a particle at radius x as if all the mass within this radius were concentrated at 0. This was discussed in Chapter 3. One can make use of this result to write down the gravitational acceleration of a particle in a spherically symmetrical distribution. It is

$$\ddot{r}_i(t) = - \frac{GM}{[r_i(t)]^2}, \quad M \equiv \frac{4\pi}{3} \rho_0 [r_i(t_0)]^3 \qquad (5.10)$$

where the total mass M *within the sphere of radius* $r_i(t)$ has been expressed in terms of the matter density (assumed uniform) at the reference time t_0. One may suppose that a cosmic repulsive force, as introduced in eqn (4.2'), acts in addition so that the generalized acceleration of particle i is

$$\ddot{r}_i(t) = - \frac{GM}{[r_i(t)]^2} + \frac{\lambda}{3} r_i(t). \qquad (5.11)$$

Eqn (5.11) is an example of an equation of motion, as discussed in connection with (4.23). It is assumed that the motion is radial.

In order to integrate (5.11), multiply the equation by $2\dot{r}_i(t)$ and treat GM and λ as constants. Hence

$$2\dot{r}_i(t)\ddot{r}_i(t) = - \frac{2GM\dot{r}_i(t)}{[r_i(t)]^2} + \frac{2}{3} \lambda \dot{r}_i(t)r_i(t).$$

Integrating each term with respect to time from some fixed time to t yields

$$[\dot{r}_i(t)]^2 = \frac{2GM}{r_i(t)} + \frac{\lambda}{3}[r_i(t)]^2 + Q_i$$

where Q_i is a constant of integration which may depend on the particle i, but is independent of time. Dividing by $[r_i(t_0)]^2$

and using eqns (4.3) and (5.10) one finds

$$[\dot{\bar{R}}]^2 = \frac{8\pi G\rho_0}{3} \frac{1}{\bar{R}} + \frac{\lambda}{3}[\bar{R}]^2 + \frac{Q_i}{[r_i(t_0)]^2}. \qquad (5.12)$$

This is again the cosmological differential eqn (4.11), provided only that one identifies Q_i as follows:

$$Q_i = \frac{E}{A}[r_i(t_0)]^2 = \frac{5[r_i(t_0)]^2 E}{2\pi\rho_0 a_0^5} \qquad (5.13)$$

where eqn (5.3) has been used. Thus the energy of the energy-
equation approach appears here essentially as a constant of
integration. Note that from eqn (5.10) there follows an
alternative way of writing eqn (5.11):

$$\ddot{\bar{R}} = -\frac{4\pi}{3} G\rho_0 \bar{R}^{-2} + \frac{\lambda}{3}\bar{R}. \qquad (5.14)$$

The analogous result based on the scaled eqn (4.15) is

$$\ddot{R} = -\frac{C}{2R^2} + \frac{1}{3}\lambda R, \qquad (5.15)$$

where C is given by eqn (5.8).

5.3. THE INFINITE UNIVERSE; THE COSMOLOGICAL FORCE

The derivation of the cosmological differential equation in
the preceding sections raises some questions of principle.
There is, firstly, a question of self-consistency. It is
assumed in Section 4.3 that Hubble's law applies at all times,
and in Section 5.1 that the universe is of uniform density in
space at all times, and it is not obvious *a priori* that these
assumptions are consistent with the detailed equations of
motion such as eqn (5.11). The second question that arises is
whether the derivation can be extended to a spatially infinite
universe, since the finite model universe considered hitherto
violates the cosmological principle of Section 2.2. The two
questions will be considered separately.

(a) *The self-consistency problem*

It will be assumed that Hubble's law applies, and that

the density $\rho(t)$ is uniform in space, at the reference time t_0. Then it is sufficient to show that Hubble's law applies at all other times t, since, as discussed in Section 2.1, this implies the existence of a universal scale factor $\bar{R}(t)$. The volume bounded by any set of particles will then vary with time as $[\bar{R}(t)]^3$, and the matter density inside that volume will vary as $[\bar{R}(t)]^{-3}$, so that an initially uniform density will remain uniform.

To discuss the validity of Hubble's law at different times, consider first the equation of motion (5.10), which assumes only that the universe is uniform at time t_0. Defining a scale factor for particle i by

$$\bar{R}_i \equiv r_i(t)/r_i(t_0), \tag{5.16}$$

eqn (5.10) can be written

$$\ddot{\bar{R}}_i r_i(t_0) = -\frac{K[r_i(t_0)]^3}{[\bar{R}_i r_i(t_0)]^2}, \quad K \equiv \frac{4}{3}\pi G\rho_0 \tag{5.17}$$

and $r_i(t_0)$ cancels to give

$$\ddot{\bar{R}}_i = -K/\bar{R}_i^2.$$

Thus $\bar{R}_i$ satisfies a differential equation which does not explicitly depend on i. Since, by assumption, Hubble's law is valid at t_0, so that $\bar{R}_i(t_0) = 1$ and $\dot{\bar{R}}_i(t_0) = H(t_0)$, we have a differential equation with parameter (K) and boundary conditions independent of i. It follows that the solution is also independent of the individual particle i considered. Hence $\bar{R}_i(t)$ is a universal scale factor $\bar{R}(t)$, and Hubble's law is valid at all times.

Suppose next that a cosmological force of the form (4.2′) with general power b is present. The equation of motion (5.11) is then modified to

$$\ddot{r}_i(t) = -\frac{GM_i}{[r_i(t)]^2} + \frac{\lambda}{3}[r_i(t)]^b$$

and the analogue of eqn (5.17) is

$$\ddot{\bar{R}}_i r_i(t_0) = - \frac{Kr_i(t_0)}{\bar{R}_i^{\,2}} + \frac{\lambda}{3}(\bar{R}_i)^b \, [r_i(t_0)]^b.$$

Evidently, cancelling $r_i(t_0)$ will give an equation for $R_i(t)$ independent of i if, and only if, $b = 1$. This is therefore a condition for the Hubble law to remain valid at all times, and it justifies the choice of $b = 1$ in eqn (4.2').

     If one regards the gravitational force as an unknown function of distance, one can similarly infer an inverse square law from Hubble's law in a Newtonian universe [1].

(b) *The problem of an infinite universe*
It is necessary here to pose one's question with precision. One can ask (i) can the approach of the previous section deal with an actually infinite universe, and (ii) can this approach deal with an arbitrarily large finite universe? The answer is 'No' to (i) and 'Yes' to (ii), for reasons given below.

     With regard to (i), an energy conservation equation for an infinite universe would consist of infinite terms and so would be indeterminate. The equation of motion (5.11) would also be indeterminate, for a reason which can be seen by reference to Fig. 5.2. In this figure we have imagined

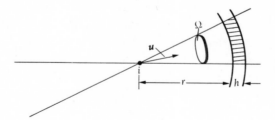

FIG. 5.2. The gravitational force on a particle is due to the mass in the intersection of a cone with i at its apex and a uniform thin concentric spherical shell at distance r from i.   This force is proportional to the solid angle Ω of the cone.

concentric spherical shells, centred on particle i, and have also drawn a narrow cone through particle i so that the gravitational force on particle i due to any particle within the

cone is roughly parallel to the fixed unit vector **u**. Then
the mass in the shaded region is, if $h \ll r$, $\rho(t)\Omega r^2 h$ where Ω
is the solid angle of the cone, and the gravitational force
due to this mass is $Gm_i\rho(t)\Omega r^2 h/r^2 = Gm_i\rho(t)\Omega h$. Adding up the
all the forces due to all such regions, and noting that half
these regions lie to the left of particle i, gives a net force
on particle i, due to all particles within the cone:

$$\mathbf{F}_i = GM_i\rho(t)\Omega h\{1-1+1-1+1-1\ldots\}\mathbf{u}.$$

Now it is known that the terms in the infinite series in
brackets can be rearranged to give any integral value whatever, for instance

$$(1-1) + (1-1) + (1-1) \ldots = 0 + 0 + 0 \ldots = 0,$$

$$1 + (-1+1) + (-1+1) + (-1+1) = 1 + 0 + 0 + 0 = 1,$$

so that no definite value can be assigned to $\mathbf{F}_i$ or, by implication, to the total force on particle i. This justifies the
conclusion above that neither the energy-conservation nor the
force approach can deal with an infinite universe.

    With regard to (ii) above, imagine two model universes
M_1 and M_2 which differ only in that the radius of M_2 at the
reference time t_0 is greater than that of $M_1 : a_2(t_0) > a_1(t_0)$.
Suppose that the two models have equal values of G, λ, ρ_0, and
H_0. The coefficients of the cosmological differential eqn
(5.12) will then be the same for M_1 and M_2, and the boundary
conditions $\bar{R}_0 = 1$ and $\dot{\bar{R}}(t_0) = H_0$ are also the same. It follows that the time variation of $\bar{R}$ is the same in both models,
and that the motion of each particle of M_1 is duplicated by
the motion of a corresponding particle in the interior of M_2.
One can therefore *replace* M_1 by M_2, or equivalently one can
increase the size of a model universe M_1 to any finite extent,
without affecting the predicted motion of any particle of M_1
or the predicted time variation of $\bar{R}$. One could obtain the
same result by simply noting that increasing the size of M_1
is equivalent to adding a uniform spherical shell, and this

shell will by eqn (3.14) exert no force on the particles of M_1.

5.4. THE FRIEDMANN MODELS

The analysis of eqn (5.9) will occupy us in Chapters 6 and 7, so that some general remarks may be helpful at this point. Various assumptions may be made concerning the values of λ and k, and these lead to various models of the smeared out universe. These models are called *Friedmann models* after the Russian mathematician who first considered them. They are now regarded as among the simplest such models and they were first investigated as a consequence of the general theory of relativity.

The first property of eqn (5.9) to which attention can be drawn is that it involves the time t or its differential dt explicitly only through the term $(\dot{R})^2$. This term is unaltered by changing the direction of the time scale since, if the new time coordinate be $t' = -t$,

$$\left(\frac{dR}{dt'}\right)^2 = \left(\frac{dR}{dt}\right)^2.$$

This equation is therefore unaltered under this transformation, and one says that it is *time-symmetrical*. It can describe an expansion from some initial value $R(0)$ at $t = 0$, as will be assumed in accordance with Hubble's law; but it can equally well describe a collapse; it can also describe an oscillation. For the moment the possibility of a collapse being described by the equations will be ignored. When one assumes that $R(0)$ is zero or small with $\dot{R} > 0$, one is considering the so-called *big-bang Friedmann models*.

5.5. WHY IS THERE NO PRESSURE TERM IN THE ENERGY EQUATION?

Some people might expect a pressure term in the energy equation. The reason why it is not there will now be given.

If we imagine a small flat smooth surface to be fixed at some point in space, the expanding systems of particles give rise to occasional collisions with this surface. We can assume that such a particle will be reflected after impact,

so that its momentum $\mathbf{p}_n$ normal to the plane will be reversed
while its momentum $\mathbf{p}_t$ tangential to the plane will remain constant. In this way a collision leads to the transfer of
momentum $2\mathbf{p}_n$ to the surface. If such collisions are allowed
to occur for a period of time, *an average rate of momentum
transfer* to the plane takes place. Since the surface is
fixed, this average rate is also the average force to be
exerted in order to keep the surface fixed in spite of the
collisions (force = rate of change of momentum!). This is
another way of saying that the particles exert a *pressure* on
the surface.

   If the small surface moves with a particle, or takes
part in the general expansion as if it were a particle, then
of course no collisions will occur. We then speak of a
comoving element of surface. On such a surface no pressure
is exerted (see also Problem (4.1)).

   Let us now consider the differential equation (5.9) for
the whole universe and note that its radius has cancelled out
of the equation, leaving a differential equation for the scale
factor. This equation is therefore appropriate for *any* comoving spherical concentric surface drawn within the universe.
Such a surface may be thought of as 'frozen into the matter'.
If such a surface were solid no particles would collide with
it since it is comoving and therefore no pressure would be
exerted on it. It is therefore plausible that the energy
equation for the matter within the surface contains no pressure terms. *In this sense the particle model of a Hubble law
universe is a zero-pressure model.* It is of course clear that
there is no pressure term in the equation for the contents of
a surface of *any* shape frozen into the matter.

   In Section 5.1 the notion of a continuum was introduced
merely as a device to calculate the coefficients in the differential equation, and one was still operating with what is
basically a particle model. If the continuum is regarded as
a limiting case of a closely spaced distribution of particles
then the above arguments still apply, and one has a zero-
pressure model, so long as each particle obeys Hubble's law
exactly. However, if random motions are superimposed on the
Hubble law expansion, then even a comoving surface will be

subject to collisions and hence one would expect pressure to play a part in the dynamics of the universe. Such models are *continuum models* in the usual sense and they will be discussed later (Chapter 9). For the moment only zero-pressure models will be considered.

PROBLEMS

(5.1) Show that $\sum_1^n r \sim n^2/2$ and $\sum_1^n r^2 \sim n^3/3$ for large n (i) by using the exact formulae, (ii) by integration.

(5.2) Show that $\sum_{r=1}^{r=n} r^\alpha \sim \dfrac{n^{\alpha+1}}{\alpha+1}$ for large n for $\alpha > 0$.

(5.3) Show that $n! \sim (n/e)^n$ for large n, where e is the basis of the natural logarithms.
[*Hint*. Consider $y = \ln n! = \sum_{r=1}^n \ln r$.]

(5.4) Show that the expression (5.7) for B may also be obtained by considering spherically symmetrical shells to condense sequentially from infinity to form uniform spheres with centre 0.

SOME SIMPLE FRIEDMANN MODELS

6.1. INTRODUCTION

The smeared out model universes have now been specified adequately, and the physical pictures and explanations on which these specifications were based have been incorporated in an energy equation (5.9) or an equation of motion (5.15). There are also boundary conditions which state whether the model is expanding or contracting at a particular time, as discussed in Section 5.4.

These procedures are illustrative of the general 'model-making' activity which has been known for so long in applied mathematics, physics, and chemistry. It has also entered the biological and social sciences. The 'model' is often in all these cases a differential equation with boundary conditions.

In this section the solution of these equations will be considered for various special cases. These cases are either physically important or they are instructive as giving insight into simpler properties of these equations. We shall then be ready to consider the more general properties of our models in Chapter 7.

6.2. MODEL UNIVERSES WITHOUT COSMOLOGICAL FORCE ($\lambda = 0$)

6.2.1. *Classification into expanding and oscillating models: the critical mass density*

The models of this section are important because they require only known forces, so that an explanation of the astronomical observations in terms of them would be particularly satisfying. In fact, many cosmologists consider that the $\lambda = 0$ models are the only ones to be used.

The key observation to be made in this case is to write our main equation (5.9) in the following form

$$kc^2 = -\dot{R}^2 + \frac{C}{R} = -R^2 H^2 + \frac{8\pi}{3} G\rho R^2$$

so that

and

$$kc^2 = (\rho - \rho_c)\frac{8\pi}{3}GR^2, \quad \rho_c \equiv \frac{3H^2}{8\pi G}$$

$$\left.\frac{\rho}{\rho_c} = \frac{C}{H^2 R^3} = \frac{C}{\dot{R}^2 R} = 2\sigma = 2q \right\}$$

(6.1)

where eqn (4.24) has been used. A critical matter density ρ_c has here been introduced. It has been related to the mass parameter σ by eqn (4.25), and it has the following inter-pretation. If at some cosmological time t one finds that $\rho > \rho_c$, then the right-hand side of eqn (6.1) is positive and so $k = +1$ at all times. Hence

$$\dot{R}^2 = \frac{C}{R} - kc^2$$

(6.2)

shows that there is a maximum scale factor $R = R_m$ which satis-fies

$$R_m = C/c^2 \quad (k = +1).$$

(6.3)

When it is reached, $\dot{R} = 0$. This maximum extent of the expan-sion is therefore succeeded by a contraction. The time re-quired to reach R_m from $R = 0$ can be evaluated by solving eqn (6.2). This is a 'detail' to which we shall turn below.

For the moment we shall consider the case that at some time $\rho < \rho_c$. Then eqn (6.1) shows that $kc^2 < 0$, and $k = -1$ at all times. In this situation eqn (6.2) shows that the model universe is an always-expanding one since $\dot{R}^2$ can never be zero.

We have therefore the main classification of $\lambda = 0$ models:

$$\left.\begin{array}{l}\rho > \rho_c \leftrightarrow q > \tfrac{1}{2} \leftrightarrow k = +1, \quad \text{an oscillation} \\ \rho = \rho_c \leftrightarrow q = \tfrac{1}{2} \leftrightarrow k = 0 \ , \quad \text{permanent expansion} \\ \rho < \rho_c \leftrightarrow q < \tfrac{1}{2} \leftrightarrow k = -1, \quad \text{permanent expansion}\end{array}\right\}$$

(6.4)

This classification is physically clear: given the value of H and therefore of ρ_c, then for large enough ρ the gravitational attraction dominates and the universe collapses again. For small enough ρ, the kinetic energy can dominate the gravitational effects and there can be indefinite expansion.

The numerical estimate $H_0 \sim 1.8 \times 10^{-18}$ s^{-1} leads to a present critical mass density

$$\left. \begin{aligned} \rho_{c0} &\sim 0.6 \times 10^{-29} \text{ g cm}^{-3} \\ &\sim 0.6 \times 10^{-26} \text{ kg m}^{-3} \sim 4 \text{ protons m}^{-3} \end{aligned} \right\} \tag{6.5}$$

The classification (6.4) also results if one assumes the density to be given. One then finds that the condition for indefinite expansion is a constraint on H, namely

$$H \geqslant \left[\frac{8\pi G \rho}{3} \right]^{\frac{1}{2}}$$

The right-hand side is a kind of escape velocity for the universe which corresponds to eqn (3.21).

If the model collapses again, then it approaches a state of infinite density as $R \to 0$, and this state is called a *singularity*. It is meaningless to ask if the *model* re-expands as there is no unique extrapolation through the singularity. The question has a meaning for the *universe*, but an answer is difficult to obtain because of our ignorance about the physical laws governing states of extremely high density. In view of this uncertainty the phrase 'oscillating universe' means that there may be only one oscillation.

6.2.2. *The Einstein—de Sitter model*

This model is defined by $k = \lambda = 0$, and is easily discussed. Eqn (6.2) now yields

$$R^{\frac{1}{2}} dR = C^{\frac{1}{2}} \, dt$$

whence

$$R^{3/2} = \frac{3}{2} C^{\frac{1}{2}} t$$

i.e.
$$R^3 = \frac{9}{4} C t^2. \tag{6.6}$$

Substituting for C from eqn (5.9), the right-hand side is $(9/4)(8\pi G\rho R^3 t^2/3)$ so that

$$6\pi G\rho t^2 = 1 \quad \text{or} \quad \rho = (6\pi G t^2)^{-1}. \tag{6.7}$$

Eqn (6.6) leads to

$$H = 2/3t, \quad q = 1/2. \tag{6.8}$$

The value of ρ can be found by putting eqn (6.8) into eqn (6.1). One recovers the critical mass density:

$$\rho = \rho_c. \tag{6.9}$$

One sees that $\rho = \rho_c$ at *all* times. This model is of particular interest because it, or a time reversal of it, occurs for all models of non-zero mass density in the limit $R \to 0$ [see Problem (6.2)].

6.2.3. *The 'missing matter' problem*

The relation between red-shift z and luminosity distance for galaxies yields estimates of q_0 (see Sections 11.2 and 11.5) with uncertainties which can be as large as ± 1. Nonetheless they impose some constraints on the possible cosmological models via eqn (4.24). As an example, we shall consider the estimate (see Section 11.5 for more details)

$$q_0 \sim 1. \tag{6.10}$$

One can use this in the equation of motion (4.24) with the continuum model interpretation (5.9) for C to find

$$\frac{\lambda}{3H^2} = \frac{4\pi G\rho}{3H^2} - q = \frac{\rho}{2\rho_c} - q = \sigma - q. \tag{6.11}$$

Taking $\rho_{c0} \sim 0.6 \times 10^{-26}$ kg m^{-3} and the luminous (galactic) mass density of the universe at present as [1]

$$\rho_0 \sim 3 \times 10^{-28} \text{ kg m}^{-3} \tag{6.12}$$

one finds

$$\frac{\rho_0}{\rho_{c0}} \sim 0.05. \tag{6.13}$$

Thus one has from eqn (6.11)

$$\lambda \sim -3H_0^2$$

so that the cosmological force is attractive and helps to slow the expansion of the universe. But this is regarded as un-likely on other grounds. A usual interpretation of these results is to assume $\lambda = 0$ and to infer from eqns (6.10) and (6.11) that $\rho_0 \sim 2\rho_{c0}$ in contradiction to eqn (6.13), since $2\rho_{c0}$ is much larger than the luminous matter density actually observed. The search for this 'missing matter' is still incom-plete, but the necessity for it may be removed completely if lower values of q_0 are accepted.

Also if the gravitational constant has had greater values in the past than it has now, one would expect a smaller mass density to have the same effects as a larger mass density has later. The result would again be that the missing matter problem is quantitatively less severe [2]. It had been proposed in the 1930s that H and G have the same time-dependences [3] and the value

$$\dot{G}/G \sim -8 \times 10^{-11} \text{ year}^{-1} \tag{6.14}$$

has recently been proposed [4,5]. Some consequences of a pos-sible time dependence of G are discussed in Appendix B.

6.2.4. *General theory of the* $\lambda = 0$ *cases*

No convincing evidence is available for a non-zero value to the cosmological constant. For this reason one puts $\lambda = 0$ in many investigations and such models therefore assume par-ticular importance. We now discuss these cases. Because of eqn (4.14) one expects oscillations for $k = +1$, since the energy is then negative, and indefinite expansion for $k = -1$, since the energy is then positive. The calculations below confirm this in detail.

Take as the main equation (5.9) in the form

$$\frac{dR}{dt} = \left(\frac{C}{R} - kc^2\right)^{\frac{1}{2}} \quad (\lambda = 0) \tag{6.15}$$

and define a new dimensionless variable u by

$$R \equiv \frac{C}{c^2}u^2 \quad (k = +1 \quad \text{or} \quad -1). \tag{6.16}$$

Since $R \geqslant 0$ by definition, u is always real, but for the case $k = +1$ is restricted to values between ± 1 by eqn (6.3). We now discuss the Hubble parameter, the deceleration parameter, and the age since the big bang in turn.

(i) *Hubble parameter*. From $\ln R = \ln C/c^2 + 2 \ln u$, and eqns (6.15) and (6.16),

$$H = \frac{\dot{R}}{R} = \frac{2\dot{u}}{u} = \frac{c^2}{Cu^2}\left(\frac{C}{R}-kc^2\right)^{\frac{1}{2}}$$

$$= \frac{c^2}{Cu^2}\left(\frac{c^2}{u^2}-kc^2\right)^{\frac{1}{2}}$$

$$= \frac{c^3/C}{u^3}\left(1-ku^2\right)^{\frac{1}{2}}.$$

Hence

$$H = \frac{2\dot{u}}{u} = \frac{\alpha}{u^3}\left(1-ku^2\right)^{\frac{1}{2}} \tag{6.17}$$

where

$$\alpha \equiv \frac{c^3}{C} .$$

The dependence of u on time is obtained below.

(ii) *The deceleration parameter*. From eqn (5.15) with $\lambda = 0$, (6.15), and (6.16),

$$q = -\frac{R\ddot{R}}{\dot{R}^2} = \frac{R(C/2R^2)}{(c/u)^2 - kc^2} = \frac{Cu^2}{2Rc^2(1-ku^2)} ,$$

so that

$$q = \frac{1}{2(1-ku^2)} \tag{6.18}$$

i.e.,

$$u^2 = k^{-1}\left(1-\frac{1}{2q}\right) = k\left(1-\frac{1}{2q}\right) \quad (k \neq 0). \tag{6.18'}$$

The case $k = 0$ will be ignored here as it has been discussed in Section 6.2.2.

   (iii) *The age*.   From eqn (6.17)

$$t = \int_0^t dt = \frac{2}{\alpha} \int_0^u \frac{u^2 du}{(1-ku^2)^{\frac{1}{2}}} \equiv \frac{f(u)}{\alpha} , \tag{6.19}$$

where

$$f(u) \equiv 2\int_0^u \frac{u^2 du}{(1-ku^2)^{\frac{1}{2}}} = \begin{cases} \sin^{-1} u - u\sqrt{(1-u^2)} & (k = +1) \\ \\ u\sqrt{(1+u^2)} - \sinh^{-1} u & (k = -1) \end{cases}$$

The evaluation of the integral is discussed in the appendix at the end of this chapter.   Clearly the present age t_0 is obtained from u_0 via eqn (6.19).   By multiplying eqns (6.17) and (6.19) one finds the age, expressed in terms of the Hubble time H^{-1}:

$$\frac{t}{t_H} \equiv Ht = u^{-3}(1-ku^2)^{\frac{1}{2}} f(u). \tag{6.20}$$

Using eqns (6.18') and (6.20), t/t_H can now be found as a function of q.   The graph of this function (Fig. 6.1) shows that the time since the big bang drops below the Hubble time as a result of deceleration.   This result is easily understood if it is recalled that a decreasing 'velocity of expansion' $\dot{R}$ implies that the function $R(t)$ is concave towards the time axis (Fig. 6.2).

   Note that the present value of q is important in two ways: it determines the type of $\lambda = 0$ universe by eqn (6.4). It also determines the present value of u by eqn (6.18').

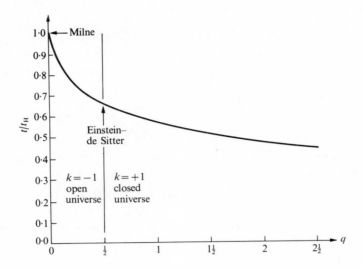

FIG. 6.1 The time since the big bang in models with $\lambda = 0$.

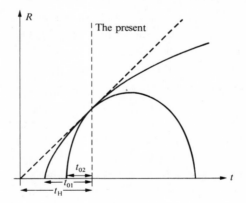

FIG. 6.2 Possible universes for $\lambda = 0$. Key for Figs. 6.1 and 6.2: t_H is the time since the big bang corresponding to a universe with $q = 0$, t_{01} is the time since the big bang corresponding to a universe with $0 < q \leqslant \frac{1}{2}$, t_{02} is the time since the big bang corresponding to a universe with $\frac{1}{2} < q$.

This yields the time since the big bang in units of the Hubble time by eqn (6.20). Thus the present value of the Hubble parameter fixes the time scale of the expansion.

The constraints on q_0 are in the main the following.
(a) Recessional velocities. These strongly suggest $q_0 < 2$ (cf.
Section 11.5). (b) The oldest stars are those of the globular
clusters, for which theories of stellar evolution suggest ages
of the order of 14×10^9 years [6]. This value yields

$$\frac{t_0}{t_H} \gtrsim \frac{14}{18} \sim 0.78, \quad \text{i.e.} \quad q_0 \lesssim 0.2$$

where the recent estimate $t_H \sim 18 \times 10^9$ years has been adopted.
(c) Since any given radioactive element decays at a constant
rate, the ratio of its abundance to the abundance of its decay
products yields the average age of the radioactive element.
Applied to heavy elements, one finds ages of the order
11.7×10^9 years [7], and the time since the big bang must be
greater than this value. In this case

$$\frac{t_0}{t_H} \gtrsim \frac{11.7}{18} \sim 0.65, \quad \text{i.e.} \quad q_0 \lesssim 0.55.$$

More general inequalities, valid for $\lambda \neq 0$, are discussed in
Appendix C. (d) The mass associated with galaxies yields densities much less than the critical density eqn (6.5), suggesting via eqn (6.11) for $\lambda = 0$ models that q_0 is much less
than 0.5. (e) The deuterium abundance is a very sensitive
indicator of the present mass density of the universe. Deuterium was formed in the first few minutes after the big bang,
and if the density then was too great, most of the deuterium
would have been turned into helium. Thus for the cosmological
element formation to leave us with enough deuterium at present,
the density of the present universe must be low enough, say
about 2×10^{-31} gm cm^{-3} [8]. This yields $q_0 \sim 0.1$. However,
all these estimates are subject to uncertainties. For example,
should it be discovered that deuterium can be produced in
supernova explosions, then the present deuterium abundance
would not be entirely due to its production in the big bang,
and q_0 could be larger, as already noted in Section 1.4.

6.2.5. *The* $k = +1$ *case*

If $k = +1$, it is easy to see (Problem (6.2) and Fig. 6.3)

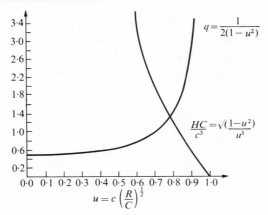

FIG. 6.3 Hubble parameter H and deceleration parameter q for a model with $\lambda = 0$ and $k = 1$. The model is capable of one oscillation and reaches its maximum extent at a time $t = (\pi/2)(C/c^3)$ after the big bang.

that soon after the big bang the Hubble parameter decreases from large values as $(2/3t)$, while the deceleration parameter is $1/2$. As R and u increase with time, H decreases further and q increases until, at $u = 1$, the scale factor R reaches a maximum value R_m given by eqn (6.3). At this time, which we denote by $t = t_m$, $H = 0$ and $q = \infty$. Since the second derivative of R is always negative for models with $\lambda = 0$ (see eqn (4.23)), the attainment of the maximum scale factor is followed at once by a contracting phase in which R decreases from R_m to zero, H increases from zero to infinity, and q decreases from infinity to $1/2$. These details follow from the time-reversibility of the equation of motion, which also implies that the time for a complete oscillation, in which R goes from zero to R_m and back to zero, is $T = 2t_m$.

Given this general picture of an oscillating universe, one may ask (i) how long is an oscillation, i.e. what is the expression for T, and (ii) how does the maximum extent of the universe compare with its present extent, i.e. what is the ratio R_m/R_0? The answer can be given in terms of the present value of either the deceleration parameter q or the density ρ, these quantities being related through eqns (6.1) and (4.24) with $\lambda = 0$:

$$q_0 = \sigma_0 = \rho_0/2\rho_{c0}. \qquad (6.21)$$

The time T is found as the first time after $t = 0$ at which $R = 0$ and hence $u = 0$. From eqn (6.19) this is

$$T = \frac{1}{\alpha} \sin^{-1}(0) = \frac{\pi}{\alpha} = \frac{\pi C}{c^3}. \qquad (6.22)$$

C can be found from eqn (4.30) with $L_0 = 0$ and with σ_0 replaced by q_0 according to eqn (6.21). The result is

$$T = \frac{2\pi q_0}{H_0} |1-2q_0|^{-3/2} = \frac{\pi q_0}{H_0 \sqrt{(2)} (q_0-0.5)^{3/2}} \qquad (6.23)$$

since $q_0 > 0.5$ for $k = +1$.

The present value R_0 of the scale factor is found from eqn (6.18') with $k = +1$ and is

$$R_0 = \frac{C}{c^2} u_0^2 = \frac{C}{c^2}\left(1-\frac{1}{2q_0}\right);$$

comparison with eqn (6.3) gives

$$R_m/R_0 = \left(1-\frac{1}{2q_0}\right)^{-1} = \frac{q_0}{q_0-0.5} \qquad (6.24)$$

as the factor by which intergalactic distances increase in going from the present time to the time of maximum scale factor.

As an example, consider the estimate $q_0 \sim 1$ introduced in Section 6.23. This leads to

$$T \sim \frac{2\pi}{H_o} \sim 11 \times 10^{10} \text{ years}; \quad R_m/R_0 \sim 2 \qquad (6.25)$$

where the estimate 1.8×10^{10} years for H_0^{-1} has also been used. Note also that as q_0 decreases towards the critical value 0.5, or equivalently as the present density ρ_0 decreases towards the critical density ρ_{c0}, both T and R_m/R_0 increase indefinitely: the universe finds it increasingly difficult to muster adequate gravitational attraction in order to contract again.

6.2.6. *The* k = -1 *case; the Milne model*

While an oscillation occurs for $k = +1$, indefinite expansion takes place for $k = -1$, as noted in eqn (6.4). The dynamical parameters are given directly by eqns (6.17) and (6.18):

$$H = \frac{c^3}{C} \frac{(1+u^2)^{\frac{1}{2}}}{u^3} \, , \tag{6.26}$$

$$q = \frac{1}{2(1+u^2)} \, . \tag{6.27}$$

It is seen that both decrease steadily as the expansion continues (Fig. 6.4).

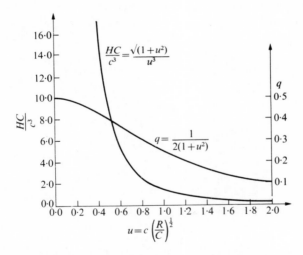

FIG. 6.4 The Hubble parameter H and deceleration parameter q for a model with $\lambda = 0$ and $k = -1$. This is an indefinitely expanding model.

After a sufficiently long time in a $(k = -1, \lambda = 0)$-model the scale factor is so large that the differential equation governing the expansion thereafter can be taken to be

$$\dot{R}^2 = c^2, \quad \text{i.e. } R = ct. \tag{6.28}$$

In this (terminal) part of the history of the universe

$$H = 1/t \text{ and } q = 0. \tag{6.29}$$

Since the deceleration parameter is zero in a model specified by eqns (6.28) and (6.29), every galaxy moves with constant speed. This is therefore the Milne model introduced in Section 2.6.

6.3. MODELS OF ZERO TOTAL ENERGY

These models are specified by $k = 0$ in Newtonian cosmology. The basic differential equation becomes

$$\dot{R}^2 = \frac{C}{R} + \frac{\lambda}{3}R^2. \tag{6.30}$$

This equation can be integrated. To see this one can change the dependent variable by means of a constant β, to be determined later, as follows:

$$y = \beta R^3, \quad \dot{y}^2 = 9\beta^2 R^4 \dot{R}^2 = 9\beta^2 R^4 \left(\frac{C}{R} + \frac{\lambda}{3}R^2\right).$$

It follows that

$$\dot{y}^2 = 3\lambda\left[\frac{3\beta C}{\lambda}y + y^2\right] = 3\lambda(2y + y^2).$$

In the last step we have chosen

$$\beta = 2\lambda/3C,$$

which conveniently removes some of the factors from the differential equation. The equation is now

$$\frac{\mathrm{d}y}{\sqrt{(2y + y^2)}} = \frac{\mathrm{d}y}{\sqrt{((y+1)^2 - 1)}} = \sqrt{3\lambda} \,\, \mathrm{d}t.$$

Integrating from $t = 0$, when it is assumed that $R = 0$, i.e. $y = 0$, one has with $z \equiv y + 1$

$$\int_0^y \frac{\mathrm{d}y}{\sqrt{((y+1)^2 - 1)}} = \sqrt{3\lambda} \,\, t \,\, \text{ or } \,\, \int_1^z \frac{\mathrm{d}z}{\sqrt{(z^2 - 1)}} = \sqrt{(3\lambda)}\,t.$$

The left-hand side is a standard integral which yields $\cosh^{-1} z$. Since $\cosh x = \frac{1}{2}(e^x + e^{-x})$, $\cosh 0 = 1$, so that the lower limit of the integral contributes $\cosh^{-1} 1 = 0$. It follows that

$$\cosh(\sqrt{(3\lambda)}t) = z = y + 1 = \frac{2\lambda}{3C}R^3 + 1.$$

Hence

$$\frac{2\lambda}{3C}R^3 = \cosh(\sqrt{(3\lambda)}t) - 1 \qquad (6.31)$$

Two cases arise in this model. If $\lambda > 0$, the model is an expanding one. Indeed for large times one approaches the condition of an almost empty universe with

$$R^3 = \frac{3C}{2\lambda} \cdot \frac{1}{2} \exp(\sqrt{(3\lambda)}t) \qquad (6.32)$$

so that

$$\left.\begin{array}{l} R(t) = \left[\dfrac{3C}{4\lambda}\right]^{1/3} \exp|\sqrt{(\lambda/3)}t| \\[3mm] H = \sqrt{(\lambda/3)} \\[3mm] q = -1. \end{array}\right\} \qquad (6.33)$$

This condition of negligible matter density is approached by many indefinitely expanding models and it is sometimes referred to as the de Sitter model.

It is a fairly simple matter to obtain the Hubble parameter and the deceleration parameter

$$H = \left(\frac{\lambda}{3}\right)^{\frac{1}{2}} \frac{\sinh \sqrt{(3\lambda)}t}{\cosh \sqrt{(3\lambda)}t - 1} \qquad (6.34)$$

$$q = \frac{2 - \cosh \sqrt{(3\lambda)}t}{1 + \cosh \sqrt{(3\lambda)}t} \ . \qquad (6.35)$$

Eqn (6.34) is found by differentiating eqn (6.31) with respect to time and multiplying the result by $(C/2\lambda R^3)$ to find

$$H = \frac{\dot{R}}{R} = \frac{C}{2\lambda R^3} \sqrt{(3\lambda)} \sinh \sqrt{(3\lambda)}t.$$

Eqn (6.34) follows by using eqn (6.31) in the result.

To obtain eqn (6.35), two differentiations of eqn (6.31) and multiplication by $C/4\lambda R\dot{R}^2$ yield

$$1 + \tfrac{1}{2} \frac{R\ddot{R}}{\dot{R}^2} = \frac{3C}{4R\dot{R}^2} \cosh(\sqrt{(3\lambda)}t).$$

On the left we have $1 - q/2$. The right-hand side involves $R\dot{R}^2$ which is, by eqns (6.30) and (6.31),

$$C\left[1 + \frac{\lambda}{3} \frac{R^3}{C}\right] = C\{1 + \tfrac{1}{2}[\cosh \sqrt{(3\lambda)}t - 1]\}$$

$$= \frac{C}{2}[1 + \cosh \sqrt{(3\lambda)}t]$$

Hence

$$1 - \frac{q}{2} = \frac{3}{2}[1 + \cosh \sqrt{(3\lambda)}t]^{-1}$$

and this yields eqn (6.35) in the form

$$q = 2 - \frac{3}{1 + \cosh \sqrt{(3\lambda)}t}.$$

In the limit $\lambda \to 0$ one can use

$$\sinh x \sim x + \frac{x^3}{3!}, \quad \cosh x \sim 1 + \frac{x^2}{2} + \frac{x^4}{4!}$$

to again find the characteristics (6.8) of the Einstein–de Sitter model.

Recall that for a real a

$$\cosh ia = \tfrac{1}{2}\left(e^{ia} + e^{-ia}\right) = \cos a.$$

It follows that if $\lambda < 0$, eqn (6.31) becomes

$$\frac{2|\lambda|}{3C}R^3 = 1 - \cos(\sqrt{(3|\lambda|)}t)$$

so that

$$R(t) = \left[\frac{3C}{2|\lambda|}\right][1 - \cos(\sqrt{(3|\lambda|)}\, t)].\qquad(6.36)$$

This is an oscillating model, and the formulae obtained in this section can be re-written for this case by making the appropriate transitions from hyperbolic to trigonometric functions.

6.4. MODEL UNIVERSES IN WHICH THE COSMOLOGICAL FORCE PREDOMINATES ($C = 0$)

The cosmological differential eqn (5.9) has a number of features which are of interest in areas beyond cosmology. In order to make this as clear as possible, we shall in this section consider a simple problem of mechanics and re-interpret it in terms of our cosmological model at the end.

Problem. A small channel allows a particle P of mass m to move on a straight line through the centre, 0, of an otherwise uniform sphere of density ρ, radius R, and surface inward acceleration g_0. Discuss the motion, assuming the particle never leaves the sphere.

Solution.

(i) *The equation of motion.*

When the particle P is at radius x from the centre, the force acting on it is due to the mass within the radius x. The mass outside this radius represents a spherically symmetric distribution. One may think of it as generated by a set of concentric spherical mass shells and these do not give rise to a force at internal points. The equation of motion is therefore

$$m\ddot{x} = -\frac{4\pi}{3}x^3 \cdot \rho \cdot G\frac{m}{x^2} = -\alpha^2 mx \qquad(6.37)$$

where

$$\alpha^2 \equiv \frac{4\pi}{3}G\rho = \frac{g_0}{R}.$$

We have used the fact that the surface inward acceleration is

$$g_0 = \frac{4\pi}{3} G\rho R.$$

The sign shows that, as x grows, the particle P is subject to an increasingly strong acceleration in the negative x-direction, i.e. towards the centre 0 of the sphere. On the other hand, when P lies on the other side of 0, so that x is negative, then $\ddot{x}$ is positive. This shows that the acceleration is in the positive x-direction, i.e. again towards 0. It grows again as $|x|$ increases. One can thus see that the particle will oscillate about 0.

Motion specified by the equation

$$\ddot{x} = -\alpha^2 x \qquad\qquad\qquad (6.38)$$

is well known in many situations and is called *simple harmonic motion*. In the present case, however, it is physically clear *why* the restoring force increases with $|x|$: more and more spherical shells of matter act to attract P as $|x|$ increases. Thus simple harmonic motion is here not assumed; it results in a mathematically simple way from the given physical situation.

(ii) *Equation of motion if particle left the sphere*

The remark in the problem that 'the particle never leaves the sphere' must now be used. For if the particle left the sphere, then there would be situations for which $x > R$, the sphere would act as if *all* its mass were concentrated at its centre and the equations of motion would be

$$m\ddot{x} = -\frac{4\pi}{3}\rho R^3 \cdot \frac{Gm}{x^2} = -\alpha^2 m \, R^3 x^{-2} \quad (x \geqslant R) \qquad (6.39)$$

$$m\ddot{x} = \alpha^2 m \, R^3 x^{-2} \qquad\qquad\qquad (x \leqslant -R) \qquad (6.40)$$

For negative x the acceleration is, by eqn (6.40), in a *positive* direction, i.e. towards 0. One sees that the attraction back to the centre weakens for $|x| > R$ as $|x|$ increases, as one would expect. It is easier to jump on a mountain than at

sea level.   In order to keep to our sign convention we need
the two equations.   For $x = R$ eqns (6.37) and (6.39) become
identical and for $x = -R$ eqns (6.37) and (6.40) become identical, as is needed for consistency.

(iii) *Solution of the equation of motion*
      The solution of a second-order differential equation con-
tains two constants, A and B say.   The equation is

$$\ddot{x} = -\alpha^2 x, \quad \alpha^2 \equiv \frac{4\pi}{3}G\rho \equiv \frac{g_0}{R}, \quad (-R < x < R) \qquad (6.41)$$

and its solution can be written

$$x = A \sin \alpha t + B \cos \alpha t$$
$$\dot{x} = \alpha (A \cos \alpha t - B \sin \alpha t).$$

In order to identify these two constants, two conditions must
be imposed.   These will be taken to be incorporated in the
following:

> The particle is released with velocity v_1 at $\Big\}$ (6.42)
> radius r_1 within the sphere.

Put mathematically, we are requiring that

$$\text{at } t = 0 : x = r_1 \text{ whence } B = r_1 \qquad (6.43)$$
$$\text{at } t = 0 : \dot{x} = v_1 \text{ whence } A = v_1/\alpha. \qquad (6.44)$$

Conditions of this type which specify a physical situation at
the beginning or end of a developing situation are called
boundary conditions.   They enable us in this case to put

$$x = (v_1/\alpha) \sin \alpha t + r_1 \cos \alpha_1 t \qquad (6.45)$$

$$\dot{x} = v_1 \cos \alpha t - \alpha r_1 \sin \alpha t. \qquad (6.46)$$

(iv) *The energy equation*
      The solution (6.45, 6.46) has the obvious property that

$$\dot{x}^2 + \alpha^2 x^2 = v_1^2 + \alpha^2 r_1^2 = \text{independent of time.} \quad (6.47)$$

Since the right-hand side depends only on constants, the expressions (6.47) are independent of time. A function of the coordinates which is constant throughout the motion may be called an *integral of the equations of motion*. The form of eqn (6.47) suggests that we are dealing with an energy equation:

$$E = \tfrac{1}{2} m\dot{x}^2 + \tfrac{1}{2} m\alpha^2 x^2 = \tfrac{1}{2} mv_1^2 + \tfrac{1}{2} m\alpha^2 r_1^2 = \text{constant.} (6.48)$$

To check that this is so requires the evaluation of the potential energy of the system.

The potential energy of the particle P at position x is obtainable as follows. The force of attraction is $-\alpha^2 mx$, and if the particle moves away from 0 (i.e. in the positive direction) it is subject to the equal and opposite force $\alpha^2 mx$ which performs work $\alpha^2 mx\, dx$. Integrating over a whole path from x_0 to x, yields the increase in potential energy in taking P from x_0 to x, and this is

$$\tfrac{1}{2}\, \alpha^2 m (x^2 - x_0^2).$$

If one takes the potential energy zero at 0, when $x = 0$, then one finds the potential energy correctly with respect to this zero only if x_0 is taken to be zero. With this sign convention, one has indeed interpreted (6.48) as an energy equation. One has also made the potential energy always positive, so that E is always positive. This contrasts with the gravitational potential energy, which was made negative.

(v) *Time for a complete oscillation under gravity along a channel passing through the centre of the Earth*

Any simple harmonic motion (6.38) has a built-in periodic time, T say, of oscillation, as has been seen intuitively under (i) above. Algebraically, one observes simply that

$$\sin \alpha t = \sin[\alpha(t + \tfrac{2\pi}{\alpha})], \quad \cos \alpha t = \cos[\alpha(t + \tfrac{2\pi}{\alpha})]$$

which shows that

$$T = \frac{2\pi}{\alpha} = 2\pi\sqrt{(R/g_0)}. \qquad (6.49)$$

Treating the Earth as uniform and spherical (both rather rough approximations), one can substitute

$$R = 6.37 \times 10^6 \text{ m}, \quad g_0 = 9.81 \text{ m s}^{-2}$$

in eqn (6.49) for the radius and the surface gravitational acceleration. One finds

$$T = 2\pi\sqrt{\left(\frac{6.37}{9.81}\right)} \times 1000 \text{ s} = 5045 \text{ s} = 84.1 \text{ minutes.} \quad (6.50)$$

The average speed for the complete circuit is

$$\frac{4 \times 6.37 \times 10^6}{5.045 \times 10^3} = 5060 \text{ m s}^{-1} = \frac{5060}{330} \text{ Mach} = 15.34 \text{ Mach.} \qquad (6.51)$$

We have here taken the speed of sound (about 330 m s^{-1} in air at $0°C$) as a convenient unit of speed.

The quantity α has the general characteristic of a spring constant. The larger α, the more rapid the oscillations and the less the maximum distance of travel from the attracting centre, if all other quantities are kept constant.

The simple theory developed above will now be applied to investigate the effect of the cosmological force alone on the dynamics of the universe.

The cosmological differential equation for a model universe of negligible mass density ($C = 0$) is, by eqn (5.9)

$$\dot{R}^2 = \frac{\lambda}{3}R^2 - kc^2 \qquad (6.52)$$

and has therefore the form (6.48), provided we change the notation as follows

$$x \rightarrow R, \quad v_1 \rightarrow c, \quad \alpha^2 \rightarrow -\frac{\lambda}{3}, \quad \frac{2}{m}E \rightarrow -kc^2. \qquad (6.53)$$

For such a model universe to be an oscillating one, α must be

real by eqn (6.45) and λ therefore negative by eqn (6.53).
Inspection of eqn (6.52) now shows that $k = -1$. The scale
factor is then subject to oscillations with period

$$T = \frac{2\pi}{\alpha} = 2\pi\sqrt{(3/|\lambda|)}, \quad \text{i.e.} \quad |\lambda|T^2 = 12\pi^2. \qquad (6.54)$$

It is now the cosmological force constant λ which has the
properties of a spring constant. One sees that the greater
$|\lambda|$, the more rapid the oscillations: λ corresponds to a
single-particle cosmological attraction in the sense of
Section 4.2. It is believed that if λ is negative then

$$|\lambda| < 3 \times 10^{-36} \text{ s}^{-2} \qquad (6.55)$$

whence

$$T > 2 \times 10^{11} \text{ years.}$$

If the zero of cosmological time is taken at the last
time of extreme compression, then $x_1 = R_1 = 0$ and the Hubble
parameter can be obtained from eqns (6.45) and (6.46):

$$x = \frac{v_1}{\alpha} \sin \alpha t \rightarrow R(t) = c\sqrt{(3/|\lambda|)} \sin(\sqrt{(|\lambda|/3)}t),$$

$$\frac{\dot{x}}{x} = \frac{\alpha}{\tan \alpha t} \rightarrow H(t) \equiv \frac{\dot{R}}{R} = \frac{\sqrt{(|\lambda|/3)}}{\tan(\sqrt{(|\lambda|/3)}t)} \qquad (6.56)$$

The deceleration parameter is similarly obtainable from

$$-\frac{x\ddot{x}}{\dot{x}^2} = \tan^2 \alpha t \rightarrow q = \tan^2(\sqrt{(|\lambda|/3)}t) \qquad (6.57)$$

The model expands from $t = 0$ to $\frac{1}{4}T \equiv (\pi/2)\sqrt{(3/|\lambda|)}$. At
that time it reaches a maximum radius

$$\sqrt{(3/|\lambda|)}c \equiv R_{max} \qquad (6.58)$$

which increases as the cosmological force weakens. During
this period, the Hubble parameter which initially decreases

from large positive values as 1/t drops to zero (see Fig.
6.5).   The deceleration parameter increases to infinity such
that

$$|H(t)|^2 \, q(t) = -\frac{\lambda}{3} \qquad (6.59)$$

remains constant throughout the period.   There then follows a
period of contraction also subject to eqn (6.59), and the state
of very small $R(t)$ recurs at $t = \frac{1}{2}T$, when there is a singularity.   During this period the Hubble parameter increases from
large and negative values to zero.   Whether or not re-expansion
occurs is a difficult matter, already referred to in Section
6.2.1.

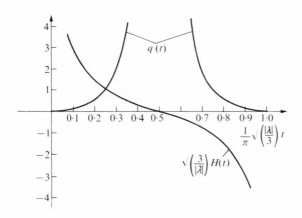

FIG. 6.5   The Hubble parameter H and deceleration parameter q for a model
with $C = 0$, $\lambda < 0$.   This is an oscillating model which reaches its maximum
extent for the first time after the initial explosion at time
$t = (\pi/2)(3/|\lambda|)^{\frac{1}{2}}$.

APPENDIX TO SECTION 6.2.4

A simple integral

With the substitution $u = \sin \vartheta$,

$$f(u) = 2\int_0^u \frac{u^2 du}{\sqrt{(1-u^2)}} = 2\int_0^\vartheta \frac{\sin^2\vartheta \cos\vartheta \, d\vartheta}{\cos\vartheta} = 2\int_0^\vartheta \sin^2\vartheta \, d\vartheta$$

$$= \int_0^\vartheta (1-\cos 2\vartheta) d\vartheta = \left| \vartheta - \frac{\sin 2\vartheta}{2} \right|_0^\vartheta$$

$$= \sin^{-1} u - u\sqrt{(1-u^2)}.$$

Similarly the substitution $u = \sinh\vartheta$ results in

$$I = 2\int_0^u \frac{u^2 du}{\sqrt{(1+u^2)}} = \left| \sinh\vartheta \cosh\vartheta - \vartheta \right|_0^\vartheta = u\sqrt{(1+u^2)} - \sinh^{-1} u.$$

PROBLEMS

On Section 6.2

(6.1)   Show that the results derived in Section 6.2.4 are in
        agreement with those which would be obtained from eqns
        (4.24) and (4.27).

(6.2)   Show that in the initial stages of a big-bang universe

$$R^3 = \frac{9C}{4}t^2, \quad H = \frac{2}{3t}, \quad q = \tfrac{1}{2}$$

        and

$$6\pi \, \rho(t)Gt^2 = 1.$$

(6.3)   Assume that we are in the early stages of a big-bang
        universe. If 'the present' corresponds to

$$t = t_0 \equiv 10^{10} \text{ years},$$

        what is the present mean density of the universe?

(6.4)   Assume that we live in a big-bang universe with $\lambda = 0$

as discussed in this Section. If $1/H_0 = 10^{10}$ years is the present value of the Hubble constant, what must be the least equivalent mass density ρ_0 at present which will ensure that the model is an oscillating one?

(6.5)   A straight channel contains a fixed particle 0 of mass M at its origin, while another particle P of mass m moves under gravitational attraction away from 0, so that its velocity tends to zero as $x \equiv OP \to \infty$.

Discuss the motion. What is the velocity of P near $x = 0$? Specify the cosmological model which has the same formal structure.

(6.6)   A straight channel contains a fixed particle of mass M at its origin 0, while another particle P of mass m moves under gravitational attraction, and has a known velocity v_0 at OP $= x_0$. Let OP be denoted by x, and let the time zero be chosen when the particle is started off from 0 in the positive x-direction.

   (i)   Show that there exists an $x = x_1$ (positive, negative, or infinity) at which the velocity of the particle vanishes and find it in terms of x_0 and v_0. Explain why x_1 is an 'effective' coordinate value in the sense that in some cases the particle will never reach this coordinate.

   (ii)   Using the substitution $x/x_1 = \sin^2(u/2)$, show that the equation of motion is with $a^2 \equiv 2GM/x_1^3$ given by $\dot{u} \sin^2(u/2) = a$, and that the total energy is $E = -GmM/x_1$.

   (iii)   Show that for $\infty > x_1 > 0$ the relation between x and the time t is

$$at = \tfrac{1}{2}(u - \sin u) = \sin^{-1}\left(\frac{x^{\frac{1}{2}}}{x_1}\right) - \left[\left(\frac{x}{x_1}\right)^{\frac{1}{2}}\left(1 - \frac{x}{x_1}\right)^{\frac{1}{2}}\right]$$

Show that the time for a complete cycle is $\tau = 2\pi/a$.

   (iv)   Show that for $0 > x_1 > -\infty$ the relation between x and t is with $v = iu = $ real

$$at = \tfrac{1}{2}(\sinh v - v) = \left[\left|\frac{x}{x_1}\right|^{\frac{1}{2}}\left(1 + \left|\frac{x}{x_1}\right|\right)^{\frac{1}{2}}\right] - \sinh^{-1}\left(\left|\frac{x}{x_1}\right|^{\frac{1}{2}}\right)$$

(v)   Interpret the above problem in terms of a model
universe with $\lambda = 0$.

On Section 6.3

(6.7)   Obtain expressions for the Hubble constant and the
deceleration parameter for a model having $k = 0$ and $\lambda < 0$.
Check that for small times they have the values appropriate to the Einstein—de Sitter model.

On Section 6.4

(6.8)   Verify that the model discussed in Section 6.4 satisfies eqns (4.24), (4.27), and (4.28).

THE CLASSIFICATION OF THE FRIEDMANN MODELS

7.1. DERIVATION OF THE CRITICAL PARAMETERS FOR THE STATIC MODEL

It is possible to write the differential equations for $\ddot{R}$ and $\dot{R}^2$ in terms of a characteristic scale factor $R*$ and a characteristic cosmological constant $\lambda*$. In order to arrive at these quantities in a natural manner we shall consider the equations $\ddot{R} = 0$ and $\dot{R} = 0$. These have to be fulfilled if the universe is static. As Einstein's original cosmological ideas of 1917 implied a static model universe, he had to introduce the cosmological constant λ *ad hoc*, and to give it the precise value $\lambda*$. The investigation of this section is therefore not only of intrinsic, but also of some historical, interest.

Two equations are needed, one for $\lambda*$ and one for $R*$ in terms of quantities already defined. The first equation comes from the requirement that

$$\ddot{R} = -\frac{C}{2R^2} + \frac{1}{3}\lambda R \qquad (5.15)$$

shall vanish when $R = R*$ and $\lambda = \lambda*$. Thus

$$2R*^3\lambda* = 3C, \text{ so that } \lambda* > 0.$$

Using this relation to replace C one finds for our two eqns (5.15) and (5.9), still kept general, but expressed in terms of $\lambda*$ and $R*$:

$$\ddot{R} = \frac{R*\lambda*}{3}\left[-\left(\frac{R*}{R}\right)^2 + \frac{\lambda}{\lambda*}\frac{R}{R*}\right]$$

$$\dot{R}^2 = \frac{R*^2\lambda*}{3}\left[2\frac{R*}{R} + \frac{\lambda}{\lambda*}\left(\frac{R}{R*}\right)^2 - 3\frac{kc^2}{R*^2\lambda*}\right].$$

We now use the fact that we want $\dot{R} = 0$ at $R = R*$. This leads to the required second equation for the $(R*,\lambda*)$-pair of constants:

$$2 + 1 - 3\frac{kc^2}{R*^2\lambda*} = 0$$

Since $\lambda* > 0$, this can be satisfied only if $k = 1$ and $\lambda* = c^2/R*^2$.

One can summarize our results as follows:

$$R* = \frac{3C}{2c^2}, \quad \lambda* = \frac{4c^6}{9C^2} = \frac{c^2}{R*^2} > 0. \tag{7.1}$$

Our two general differential equations can be written:

$$\ddot{R} = \frac{2c^4}{9C}\left[- \left(\frac{R*}{R}\right)^2 + \frac{\lambda}{\lambda*}\frac{R}{R*}\right]$$

$$\dot{R}^2 = \frac{c^2}{3}\left[2\frac{R*}{R} + \frac{\lambda}{\lambda*}\left(\frac{R}{R*}\right)^2 - 3k\right],$$

and if we now introduce a dimensionless scale factor and cosmological constant

$$x \equiv R/R*, \quad \alpha \equiv \lambda/\lambda* \tag{7.2}$$

the equations become

$$\dot{x}^2 = \beta\left\{\frac{2}{x} + \alpha x^2 - 3k\right\} \equiv \beta y \tag{7.3}$$

$$\ddot{x} = \beta\left\{\frac{-1}{x^2} + \alpha x\right\} \equiv \beta w \tag{7.4}$$

where $\beta \equiv 4c^6/27C^2$ and y, w are defined as functions of x, α, k by eqns (7.3) and (7.4).

Eqns (7.1) to (7.4) contain in a convenient form all the information needed for a complete classification of Friedmann models. The major distinction to be made is between 'oscillating' and 'continuously expanding' model universes. Two remarks are in order here. First, since (7.3) does not determine the sign of $\dot{x}$, it does not distinguish between expanding and contracting models. We choose to consider only models which are expanding for at least some part of their history, and therefore make the convention that

A model for which $\dot{x}^2$ is never zero is said to be continually expanding. (7.5)

Second, oscillating models such as that of Section 6.2.5, in which the scale factor passes through a maximum, have $\dot{R}^2 = 0$ and therefore $\dot{x}^2 = 0$ at the time of maximum expansion. However, $\dot{x}^2 = 0$ at time t_m, say, is consistent with another possibility — that the scale factor decreases from large values to a minimum at time t_m and increases thereafter. It is a matter of taste whether such a model should be described as oscillating, since there seems no way in which it can re-contract and repeat the cycle. In order to keep our classification simple we make the convention that

A model in which $\dot{x}^2$ becomes zero at any time is said to be 'oscillating'. (7.6)

The Hubble parameter and deceleration parameter can be found from eqns (7.3) and (7.4):

$$H = \frac{\dot{R}}{R} = \frac{\dot{x}}{x} = \frac{\beta^{1/2}}{x^{3/2}} \left\{ 2 + \alpha x^3 - 3kx \right\}^{1/2}, \qquad (7.7)$$

$$q = -\frac{R\ddot{R}}{\dot{R}^2} = -\frac{x\ddot{x}}{\dot{x}^2} = -\frac{xw}{y}$$

$$= \frac{1 - \alpha x^3}{2 + \alpha x^3 - 3kx}. \qquad (7.8)$$

Note that since $\dot{x}^2 \geqslant 0$ the deceleration parameter has the sign of $(1-\alpha x^3)$.

7.2. OSCILLATING AND CONTINUOUSLY EXPANDING MODELS

The condition for an 'oscillating' model is, by eqns (7.3) and (7.6), that y becomes zero. Now $y \to +\infty$ as $x \to 0$, and y has the sign of α for large enough x. Thus if $\alpha < 0$ (i.e. $\lambda < 0$) y

changes sign and therefore has at least one zero. Furthermore,

$$\frac{dy}{dx} = - \frac{2}{x^2} + 2\alpha x = 2x\left\{\alpha - \frac{1}{x^3}\right\}$$ (7.9)

implies that if $\alpha < 0$, $dy/dx < 0$ for all x. For each value of
k there is therefore just one zero of y, at $x = x_m$, say, and
$y \geqslant 0$ only for $x \leqslant x_m$ (see Fig. 7.1(a)). It follows that

Models with $\lambda < 0$ are 'oscillating' for k = - 1, 0, 1;
their scale factor goes through a maximum.   (7.10)

If $\alpha > 0$ it is evident from eqn (7.3) that y is large both
for small and for large x, and from (7.9) that $dy/dx = 0$ at
just one value of x, x_1, say. This value of x must correspond
to a minimum value of y (see Fig. 7.1(b,c)). So we define

$$x_1 \equiv \alpha^{-1/3}, \ y_{min} = \frac{2}{x_1} + \alpha x_1^2 - 3k = 3(\alpha^{1/3} - k).$$ (7.11)

If $\alpha > 1$ (i.e. $\lambda > \lambda^*$) then $y_{min} > 0$, so $\dot{x}^2 > 0$ for all
values of x, so that

Models with $\lambda > \lambda^*$ are continuously expanding for
k = - 1, 0, 1.   (7.12)

If $0 < \alpha < 1$ ($0 < \lambda < \lambda^*$) eqn (7.11) shows that $y_{min} > 0$ for
$k = 0$ and -1, and $y_{min} < 0$ for $k = +1$ (see also Fig. 7.1(c)).
It follows that $\dot{x}$ is never zero for $k = 0$ and -1, but is zero,

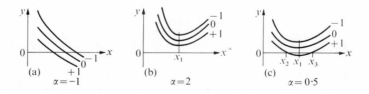

(a) $\alpha = -1$ (b) $\alpha = 2$ (c) $\alpha = 0 \cdot 5$

FIG. 7.1 The function $y(x)$ defined by (7.3), for various values of k and
α. Only non-negative values of y represent attainable states of a model
universe.

for some values of x, for k = +1, so that

> *Models with $0 \leqslant \lambda < \lambda^*$ are continuously expanding for*
> k = 0 *and* -1, *and are 'oscillating' for* k = +1.   (7.13)

The λ = 0 models have been included in (7.13) by referring
back to eqn (6.4).

The 'oscillating' models with $0 < \lambda < \lambda^*$ and k = +1 are of
two types. Since $y_{min} < 0$ but y is positive for both small and
large x, y must have two zeros, x_2 and x_3 (see Fig. 7.1(c)),
such that $\dot{x}^2 > 0$ for $x < x_2$ and for $x > x_3$. This includes (i) a
model in which the scale factor increases from zero to a maximum value $R^* x_2$, and then decreases to zero. It also includes
(ii) a model in which the scale factor *decreases* from large
values to a minimum value $R^* x_3$ and then increases without
limit. Model (i) is not qualitatively different from the
'oscillating' models of (7.10). Model (ii), however, is of
special interest as the *first Friedmann model we have encountered which is never of infinite density and therefore has no
singularities*.

The only models not included in the classification of
eqns (7.10), (7.12), and (7.13) are those with $\lambda = \lambda^*$ ($\alpha = 1$).
By eqn (7.11), these have a minimum value of y given by

$$y_{min} = 3(1-k).$$

For k = 0 and -1 these models therefore have positive y_{min} and
are expanding. For k = +1, however, y_{min} is zero, and this
zero corresponds to the Einstein static universe ($x = \alpha = k =$
1 and so from eqns (7.3) and (7.4) $\dot{x} = \ddot{x} = 0$). Near x = 1 we
can show that $x \sim 1 + (x(t_1)-1) \exp (\pm\sqrt{(3\beta)}(t-t_1))$ where t_1 is
a reference time (see Problem 7.3). Two possibilities are of
interest. If $x(t_1) > 1$ and the + sign is taken, we have a
model which was close to the static model (x = 1) in the distant past, and is expanding away from it. This, the Eddington–
Lemaître model of 1930–31, is the *second Friedmann model without singularities*, and we shall not find any others. If on the
other hand $x(t_1) < 1$ and the − sign is taken, we have a model

expanding from a zero scale factor and approaching, but never
reaching, the static model.   These results can be summarized as
follows.

> *Models with* $\lambda = \lambda^*$ *are continuously expanding for*
> k = +1, 0, *and* -1.   *For* k = +1 *the static value* R*
> *of the scale factor acts as a boundary which cannot*
> *be crossed during the expansion.*   (7.14)

7.3. THE SHAPE OF THE $R(t)$ CURVE FOR DIFFERENT MODELS

In order to discuss the shape of the $R(t)$ graph we consider
the second derivative $\ddot{x}$ given by eqn (7.4).   Evidently $\ddot{x}$ is
negative for small enough x, whatever the value of α.   Also
$\ddot{x}$ is negative for all x if α is zero or negative, while if α
is positive $\ddot{x}$ becomes positive for $x > x_1$, where $x_1 \equiv \alpha^{-1/3}$ was
introduced in eqn (7.11).   Since $\ddot{x}$ determines the curvature of
the $x(t)$ graph, it is now possible to sketch this graph for
different models.   Two main classes emerge:

(a) *Big-bang 'oscillating' models*
      These occur for $\lambda < 0$ (see eqn (7.10)), and for
$0 \leqslant \lambda < \lambda^*$ and k = +1 (see eqn (7.13)).   In all cases $\ddot{x} < 0$, so
the graph looks like Fig. 7.2(a).   The deceleration parameter
q has the sign of $(-\ddot{x})$ and so is positive.

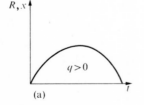
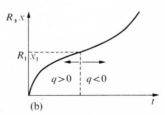

FIG. 7.2   Scale factor as a function of time for two classes of model.
(a) Big-bang oscillating models.   (b) Models with $\lambda > 0$ which expand
indefinitely.

(b)   *Big-bang models with* $\lambda > 0$ *which expand·indefinitely*
      These occur for $\lambda > \lambda^*$ (see eqn (7.12)) and also for

$0 < \lambda < \lambda^*$ if $k = 0$ or -1 (see eqn (7.13)). In these models $\ddot{x}$ changes sign at $R = R^* x_1 \equiv R_1$, and the expansion rate also goes through a minimum at this value of R (see eqn (7.11)). The graph looks like Fig. 7.2(b), and the deceleration parameter changes sign from positive to negative as R passes R_1.

7.4. MODELS EXHIBITING A MINIMUM IN THEIR EXPANSION RATES

The requirement $\dot{R} = \ddot{R} = 0$ leads to the static Einstein model, whereas the less restrictive requirement that $\dot{R} = 0$ at some time leads to the 'oscillating' models. We now consider models which have $\ddot{R} = 0$ at some time, and which therefore exhibit an extremum in their expansion rate $\dot{R}$. In fact, only a minimum in the expansion rate is possible. The condition for this to occur is

$$\ddot{x} = 0 \quad \text{i.e. } x = x_1 \text{ where } \alpha x_1^{\,3} = 1. \tag{7.15}$$

The minimum rate is obtained by putting eqn (7.15) into eqn (7.3) to find

$$(\dot{x}^2)_{\min} = 3\beta\left(\frac{1}{x_1} - k\right). \tag{7.16}$$

A minimum exists for $k = 1, 0, -1$, but it can be zero only if $x_1 = 1$ and $k = 1$. The second class of models considered in the preceding section is of this type. These models are of interest in that their age can exceed the Hubble time, as is shown in Fig. 7.3. Indeed if $k = 1$ the age can for suitably chosen $\lambda > \lambda^*$ be arbitrarily long. This will now be investigated (see Fig. 7.4).

Consider a small departure in x from $x = x_1$ given by eqn (7.15). For example let $x_2 = x_1 + \Delta$, $\Delta \ll x_1$. Then the expansion rate at this scale factor is given by eqn (7.3), if the cubic term in Δ be neglected:

$$(\dot{x}^2)_2 = 3\beta\left(\frac{1}{x_1} - 1 + \frac{\Delta^2}{x_1^{\,3}}\right). \tag{7.17}$$

This expression reduces correctly to eqn (7.16) when $\Delta = 0$. It also shows that the same expansion rate will be found at

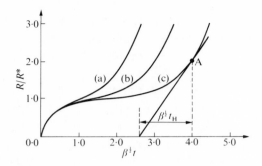

FIG. 7.3 The Lemaître model (k = +1) exhibiting increasing coasting periods as λ approaches λ^*. (a) λ/λ^* = 1.2, (b) λ/λ^* = 1.05, (c) λ/λ^* = 1.01. At point A the Hubble time divided by the age is only 1.4/4 = 0.35.

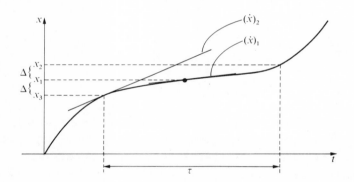

FIG. 7.4 A model with a minimum in its expansion rate.

$x_3 = x_1 - \Delta$. For $x_1 \sim 1$, i.e. for $\alpha \sim 1$, the expansion rate is thus seen to be very small in the scale factor range $(x_1-\Delta,\ x_1+\Delta)$. But the corresponding time interval τ need not be small as will now be shown.

Fig. 7.4 shows that $2\Delta/\tau$ is an underestimate of $(\dot{x})_2$ and an overestimate for $(\dot{x})_1$. Hence τ is encapsulated within two inequalities:

$$\frac{2\Delta}{(\dot{x})_2} < \tau < \frac{2\Delta}{(\dot{x})_1}$$

This tells us that

$$\frac{4\Delta^2}{3\beta} \frac{x_1^3}{x_1^2 - x_1^3 + \Delta^2} < \tau^2 < \frac{4\Delta^2}{3\beta} \frac{x_1}{1 - x_1} , \qquad (7.18)$$

so that $\tau^2 > 4/3\beta$ if $x_1 \sim \alpha \sim 1$. It shows that for $x_1 = 1 - \eta$ ($\eta \ll 1$) the minimum rate $(\dot{x}^2)_1 = 3\beta\eta$, which is near zero, can effectively continue for an interval of time which need not vanish with Δ.

During this 'coasting' period the scale factor lingers near the value $R^* x_1$. Models with a long coasting period were first considered by Lemaître. Apart from allowing an age greater than t_H, and thus avoiding conflicts of evidence of the kind mentioned in Section 1.3, they have another property of interest. According to general relativity, as noted in Section 10.7, the observed red-shift of a source depends only on the ratio of the scale factors at the times of emission and observation of the light by which the source is seen. Thus, in a universe in which the scale factor is roughly constant for a long period, many sources should be observed with red-shifts in a single comparatively small range. Quasi-stellar objects were at one time thought to exhibit such red-shift clustering, but the evidence is not conclusive [1,2].

The two-fold classification of models in this and the last section has left out several important models, those with $\lambda = 0$ or $\lambda = \lambda^*$, and also the singularity-free oscillating model of eqn (7.13). Graphs for these models are included in Fig. 7.5, which summarizes the main results of this chapter. The figure shows that indefinite expansion is encouraged by either increasing λ or by decreasing k through its permitted values $k = +1$, $k = 0$, and $k = -1$. This can be understood intuitively since the cosmological force is increasingly repulsive as λ increases, and since, for given ρ_0 and λ, the decrease of k represents an increase in kinetic energy. Both tendencies encourage expansion. Note also that the singularity-free models, like the steady-state model, find difficulty in explaining the microwave background radiation.

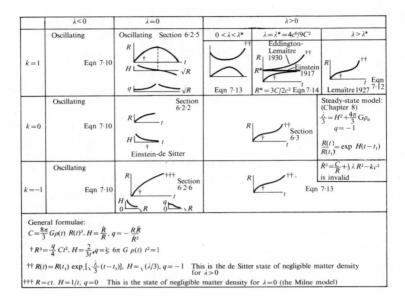

FIG. 7.5   The main characteristics of Friedmann models.

Problems

(7.1)   How do $R(t)$, $H(t)$, $q(t)$ behave after long times in the Eddington–Lemaître model ($R > R^*$, $\lambda = \lambda^*$)?

(7.2)   How do $R(t)$, $H(t)$, $q(t)$ behave soon after the big bang in the ($R < R^*$, $\lambda = \lambda^*$)-model?

(7.3)   By putting $x = (1+\vartheta)$, $|\vartheta| \ll 1$, in eqns (7.3) and (7.4), show that a small perturbation of the Einstein static model ($k = +1$, $\lambda = \lambda^*$) leads to an exponential rise or an exponential decay of x with time. Discuss in what sense this implies an instability of the Einstein static model.

8

THE STEADY-STATE MODEL

8.1. BASIC EQUATION

In this model it is assumed that the Hubble parameter, which has so far been treated as a function of time, is a true constant. The universe is in this case assumed to be in a steady state in spite of the expansion. This means that matter creation has to be postulated throughout space in order for the average matter density to remain constant in any volume element. It follows at once that the energy equation, which is the basis of the cosmological differential eqn (5.9), cannot be taken over as it stands, since it does not contain a term which allows for matter creation. However, the equation of motion (5.14) is satisfactory as it can be regarded as derived from the force acting on a particle (see eqn (5.11)), and this force need not be altered by the creation of matter. We proceed to consider the consequences of

(i) the equation of motion

$$\ddot{\bar{R}} = - \frac{\bar{C}}{2\bar{R}^2} + \frac{1}{3}\lambda\bar{R} \ , \tag{8.1}$$

where

$$\bar{C} \equiv \frac{8\pi G}{3}\rho(t)\bar{R}^3(t).$$

(ii) the steady-state conditions

$$\rho(t) = \rho_0, \ H(t) = H_0. \tag{8.2}$$

Not $\bar{C}$, but only $\bar{C}/\bar{R}^3$, is now a constant since $\rho(t)$ must be constant at ρ_0.

From the definition of the Hubble constant we have for any pair of times t, t_0 (cf. eqn (2.10)):

$$\dot{\bar{R}}(t)/\bar{R}(t) = H_0, \quad \text{i.e.} \quad \bar{R}(t) = \bar{R}(t_0)\exp\{H_0(t-t_0)\}. \qquad (8.3)$$

It follows at once that

$$\ddot{\bar{R}}(t) = H_0^2 \bar{R}(t), \quad q = -\frac{\ddot{\bar{R}}\bar{R}}{(\dot{\bar{R}})^2} = -1. \qquad (8.4)$$

Combination of eqns (8.1) and (8.4) yields an identification of the Hubble constant

$$H_0^2 = -\frac{\bar{C}}{2\bar{R}^3} + \frac{1}{3}\lambda = \frac{1}{3}\lambda - \frac{4\pi}{3}G\rho_0. \qquad (8.5)$$

This makes sense only if the right-hand side is positive.

An equation of this type was derived from a model which assumes the proton charge to be slightly in excess of the electron charge [1]. This theory was a generalization of the original steady-state theory [2] and provided a physical interpretation of the constant λ. In this theory λ is not an *ad hoc* constant for a cosmological force, but arises from the assumed charge excess. It will therefore be briefly discussed at the end of this chapter, even though it is not in agreement with experiment.

8.2. THE MATTER CREATION RATE

In order that the density of matter be constant in time a creation rate g of matter per unit volume is needed. In order to calculate it, assume radial expansion of a volume $V(t)$ in time Δt to $V(t+\Delta t)$. The volume expansion is best specified by imagining a smooth surface drawn through a set of named galaxies and thus 'frozen into the matter'. Then if this surface is a sphere whose radius increases from r to $r + h$ with $h \ll r$, then

$$V(t+\Delta t) = V(t) + 4\pi r^2 h = V(t) + 4\pi r^3 H_0 \, \Delta t \equiv V(t) + \Delta V.$$

This newly 'created' volume must be filled by galaxies of average matter density ρ_0 so that a mass

$$\rho_0 \, \Delta V = 4\pi r^3 H_0 \, \rho_0 \, \Delta t$$

is needed.   Dividing by V and Δt to obtain the creation rate
per unit volume per unit time, one finds

$$g = 3H_0 \, \rho_0. \tag{8.6}$$

Taking H_0 as about 10^{-10} years^{-1} or $\frac{1}{3} \times 10^{-17}$ s^{-1} and ρ_0
as 10^{-26} kg m^{-3} (10^{-29} gm cm^{-3}), one finds

$$g = 3 \times \frac{1}{3} \times 10^{-17} \times 10^{-26} \text{ kg m}^{-3} \text{ s}^{-1}$$

$$= 10^{-43} \text{ kg m}^{-3} \text{ s}^{-1}. \tag{8.7}$$

This is equivalent to creating one hydrogen atom per century
in a cube of side approximately 100 m since

$$\frac{2 \times 10^{-27} \text{(kg)}}{4 \times 10^9 \text{(s)} \times 10^6 \text{(m}^3)} \sim 5 \times 10^{-43} \text{ kg m}^{-3} \text{ s}^{-1}.$$

It would be extremely difficult to detect such creation rates.
    To see the error which we have avoided by using the
equation of motion, instead of the energy equation, multiply
eqn (8.5) by $\bar{R}^2$ to find

$$\dot{\bar{R}}^2 = -\frac{4\pi}{3} G\rho_0 \, \bar{R}^3 \frac{1}{\bar{R}} + \frac{1}{3}\lambda \, \bar{R}^2. \tag{8.8}$$

The energy equation would have yielded the result

$$\dot{\bar{R}}^2 = \frac{8\pi}{3} G\rho_0 \, \bar{R}^2 + \frac{1}{3}\lambda \, \bar{R}^2 + E/A, \tag{8.9}$$

and this is seen to be invalid in this case so long as E/A is
a constant.

8.3. THE THEORY OF AN EXCESS PROTONIC CHARGE

We now indicate how the force term λR can arise as a result of unequal
charges of protons and electrons in a steady-state model.   Let m_p be the
mass of a proton and $(1+y)e$ its electric charge.   The mass and charge of
an electron are m_E and $e_E = -e$.   Consider a mass M of N hydrogen atoms

$$M = Nm_p + Nm_E \sim Nm_p$$

in the form of a sphere of radius r within a concentric larger sphere. The electrostatic repulsion on a hydrogen atom situated on the outer surface of the inner sphere, due to the protonic charge excess, is (in cgs units)

$$F_E = N\frac{ye \cdot ye}{r^2} = \frac{My^2 e^2}{m_p r^2} .$$

The charge outside the sphere has no effect for a uniform distribution. The same holds for the gravitational effect. The gravitational attraction of M on the surface atom is

$$F_G = G\frac{Mm_p}{r^2} .$$

The ratio is (in cgs units)

$$\frac{F_E}{F_G} = \frac{y^2 e^2}{Gm_p^2} = \left(\frac{ye}{\sqrt{G}m_p}\right)^2 \equiv \mu . \tag{8.10}$$

The net repulsive force is, with $M \equiv 4\pi\rho_0 R^3/3$,

$$m_p\ddot{r} = F_E - F_G = F_G(\mu-1) = (\mu-1)\frac{GMm_p}{r^2} = \frac{4\pi}{3}(\mu-1)Gm_p\rho_0 r. \tag{8.11}$$

Since the particle is participating in the Hubble law expansion, and putting $r = Hr_1$,

$$\ddot{r} = \ddot{H}r_1 = H_0^2 \bar{R}r_1 = H_0^2 r, \tag{8.12}$$

where eqn (8.4) has been used. From eqns (8.11) and (8.12)

$$H_0^2 = \frac{4\pi}{3}(\mu-1)G\rho_0, \quad \text{i.e. } \mu = 1 + \frac{3H_0^2}{4\pi G\rho_0} . \tag{8.13}$$

Adopting cgs units and $\rho_0 \sim 10^{-29}$ g cm^{-3}, $H_0 \sim \frac{1}{3} 10^{-17}$ s^{-1}, eqn (8.13) gives

$$\mu \sim 1 + \frac{10^{-34}}{3\times 4\pi\times(6.67\times10^{-8})(10^{-29})} \sim 5. \tag{8.14}$$

Hence eqn (8.10) yields in cgs units[†]

$$y \sim \frac{\sqrt{(G\mu)} m_p}{e} = \frac{\sqrt{(6.67 \times 5)} \; 10^{-4} \times (1.672 \times 10^{-24})}{4.802 \times 10^{-10}} \sim 2 \times 10^{-18}. \quad (8.15)$$

The theory therefore suggests that the expansion of the universe is driven by a disparity of charge given by eqn (8.14). Indeed, comparing eqns (8.5) and (8.13), the electrostatic repulsion is seen to be equivalent to the existence of a cosmological constant

$$\lambda = 4\pi \; \mu G \rho_0. \quad (8.16)$$

In a model which is not in a steady state one would have $M = (4\pi/3) \rho(t) R^3$ instead of $(4\pi/3) \rho_0 R^3$, and would then not find an equivalent cosmological constant (8.16).

The theory in this simple form can be rejected on two conclusive grounds. First, by eqn (8.10), the ratio of electrostatic to gravitational forces F_E/F_G is the same for all spheres of hydrogen, including stars. The observed fact that stars are stable against their own radiation pressure then implies that $F_E < F_G$ and $\mu < 1$, in contradiction to eqn (8.14). Second, experiments suggest a limit on y which is much lower than eqn (8.15). In fact $y < 2 \times 10^{-21}$, leading to $\mu < 5 \times 10^{-6}$ [3].

---

† In SI units we have $e = 1.602 \times 10^{-19}$ C and

$$y = \frac{\sqrt{(66.7 \times 4.98)} \, 10^{-6} \; (1.672 \times 10^{-27})}{1.602 \times 10^{-19}/\sqrt{(4\pi\varepsilon_0)}} = 2.01 \times 10^{-18}$$

where $\varepsilon_0 = 8.85 \times 10^{-12}$ Fm^{-1}. The square root term in the denominator arises from the force law in SI units, which is $e^2/4\pi\varepsilon_0 r^2$.

UNIVERSES WITH PRESSURE

9.1. AN ENERGY CONSERVATION LAW INVOLVING PRESSURE

The interesting and important point to be made now, and to be discussed in more detail in Section 9.2, is that for uniform cosmological models we have not gone wrong in deriving eqn (5.9) or (5.15) from Newtonian mechanics, for these results follow also from general relativity and eqn (5.9) holds for non-zero pressure even though the pressure does not occur explicitly. An independent relation is therefore needed to calculate the pressure. It is given below as eqn (9.3).

The generalization of eqn (5.15) which allows for a non-zero pressure p is

$$\ddot{R} = - \frac{4\pi}{3}G(\rho+3p\mu/c^2)R + \frac{1}{3}\lambda R \tag{9.1}$$

The first term is exactly the same as the first term in eqn (5.15):

$$- \frac{8\pi}{3}G\rho[R(t)]^3/2R^2 = - C/2R^2 . \tag{9.2}$$

The last term is also the same. The second term with $\mu = 1$ arises from the general theory of relativity, when the relativistic field equations are applied to cosmological models. This term asserts that the gravitational acceleration acting on a volume element is due to the material density ρ *to which a term proportional to the pressure in the fluid has to be added*. This effect is closely related to the two relativistic notions (i) that all energy has mass, and (ii) that all mass is subject to gravitational effects. The energy involved in this case is the term $p\delta V$, the pressure energy in a volume δV, with which a material density $3p/c^2$ is then associated. In the limit $c\to\infty$ relativistic results tend to lead to corresponding results in Newtonian mechanics. Hence one can see that if the Newtonian picture is taken seriously we might equally well put $\mu = 0$, and arrive again at eqn (5.15). Thus the value of μ distinguishes between two different theories. However, only the case $\mu = 1$ (general relativity) yields results which, for the models considered here, would currently be regarded as correct.

The modification given above to Newtonian theory brought about by putting $\mu = 1$ does not involve the full apparatus of general relativity. One may think of it as New- tonian, amended by the principles (i) and (ii) discussed above. It will be termed the 'amended Newtonian theory', and we shall proceed to develop it further, and drop now the simple purely Newtonian theory. If in this relativistic (i.e. generalized) theory the pressure term is neglected, one speaks of *dust* or *incoherent* matter.

Although the matter density ρ has been assumed constant in space at any one time, it has been also assumed to change in time t. Thus changes dE in energy E within a fictitious surface S^* which is frozen into the matter can occur in time intervals dt. The law of energy conservation then requires that

$$\frac{\mathrm{d}E}{\mathrm{d}t} + p\frac{\mathrm{d}V}{\mathrm{d}t} = 0, \tag{9.3}$$

where in our amended Newtonian theory one can put $E/V = \rho c^2$ by virtue of (i). Eqn (9.3) attributes the rate of energy gain to the rate of compressional work done on the matter within the surface S^* by the fluid external to S^*. The volume within S^* has been denoted by V. The result (9.3) holds for all *uniform* cosmological models also in the general theory of rela- tivity [1], and is the independent result needed for the cal- culation of the pressure.

In addition to eqns (9.1) and (9.3), one requires an equation of state for the cosmological fluid. This relates its density to its pressure. Several examples will be given in Section 9.3.

It will also be assumed that a constant relates the volume V of S^* and the cube of the radius of S^*; this is ob- viously valid in Newtonian theory, but it also holds in

general relativity for all cases of interest in this book
(i.e. for Robertson—Walker metrics):

$$V = aR^3.$$ (9.4)

Eqns (9.3) and (9.4) imply that

$$\frac{d(\rho R^3)}{dt} = - \frac{p}{c^2} \frac{dR^3}{dt}$$ (9.5)

where ρ is the mass density and ρc^2 the energy density. This
is an equation for the pressure, a dot denoting differentiation with respect to time:

$$- \frac{1}{c^2} p = \rho + \frac{1}{3} R \dot{\rho} / \dot{R},$$ (9.6)

or, written differently,

$$\frac{\dot{\rho}}{\rho + p/c^2} = -3 \frac{\dot{R}}{R}.$$ (9.7)

This result, which (using eqn (9.4)) is equivalent to the
energy conservation eqn (9.3), holds for homogeneous models
also in general relativity, not only in the amended Newtonian
theory.

An additional assumption involved in eqn (9.3) is that
the processes which take place are slow enough for the pressure p to be representative of thermodynamic equilibrium. If
it differed from the equilibrium values the right-hand side of
eqn (9.3) would not be zero. For a change which takes place
over a period of time, this means that at each instant during
this change the system is assumed to be in thermodynamic
equilibrium. Such changes are studied in the science of
thermodynamics and are then called *quasistatic* changes [2].
In addition it is assumed that heat exchanges between different volume elements can be neglected.

For zero pressure one finds from equns (9.3), (9.5), or
(9.7)

$$\rho(t)[R(t)]^3 = \text{independent of time, i.e. } \frac{\dot{\rho}}{\rho} = - 3 \frac{\dot{R}}{R},$$ (9.8)

which we know already from our zero-pressure Newtonian model (cf. eqn (5.4)).

The first and second law of thermodynamics for fluids subject to quasistatic processes can be expressed by

$$dE + pdV = TdS, \tag{9.9}$$

a generalization of eqn (9.3). Here T is the absolute temperature and S is the entropy. This introduces a term TdS/dt on the right-hand side of eqn (9.5), a term $-T\dot{S}/3R^2\dot{R}$ on the right-hand side of eqn (9.6), and modifies eqn (9.7) to

$$\frac{1}{3}\frac{R}{\dot{R}}\frac{\dot{\rho}}{\rho+p/c^2} - \frac{T\dot{S}}{3R^2\dot{R}(\rho+p/c^2)} = 1. \tag{9.10}$$

However, the terms involving dS or $\dot{S}$ vanish for zero heat exchanges, as usually assumed in this kind of work. This is forced on the present investigations by the assumption of the isotropy of the model. Heat exchanges pick out preferred directions and so disturb isotropy.

9.2. THE DIFFERENCE BETWEEN THE AMENDED NEWTONIAN AND THE RELATIVISTIC THEORY OF UNIFORM MODEL UNIVERSES

It was seen in Section 5.2 that an equation of motion can yield an energy equation by integration. The generalized equation of motion (9.1) should therefore also yield an energy equation. To see this, multiply this equation by $2\dot{R}$ and eliminate the pressure by eqn (9.6), thus using the amended Newtonian theory, to find

$$2\dot{R}\ddot{R} = -\frac{8\pi G}{3}[(1-3\mu)\rho R\dot{R}-\mu R^2\dot{\rho}] + \frac{2}{3}\lambda R\dot{R}.$$

Integrating from a fixed time t_1 (say) such that $R(t_1) = R_1$ to a general time,

$$\dot{R}^2 = \frac{8\pi G}{3}\int_{t_1}^{t}[\mu R^2\dot{\rho}+(3\mu-1)\rho R\dot{R}]dt + \frac{1}{3}\lambda R^2 + (\dot{R}^2)_{t_1} - \frac{1}{3}\lambda R_1^2 \tag{9.11}$$

For $\mu = 1$ (i.e. for the case of general relativity) the integral in eqn (9.11) is

$$\int_{t_1}^{t} d(\rho R^2) = \rho R^2 - (\rho R^2)_{t_1}$$

so that

$$\dot{R}^2 = \frac{8\pi G}{3}\rho R^2 + \frac{1}{3}\lambda R^2 + \{\dot{R}^2 - \frac{1}{3}\lambda R^2 - \frac{8\pi G}{3}\rho R^2\}_{t_1}. \qquad (9.12)$$

If the constant of integration is denoted by

$$-kc^2 \equiv (\dot{R}^2)_{t_1} - \frac{1}{3}\lambda R_1^2 - \frac{8\pi G}{3}(\rho R^2)_{t_1}, \qquad (9.13)$$

one can see that we have recovered precisely the earlier zero-pressure Newtonian eqn (5.9), *whose range of validity has thus been enormously broadened*: it is valid for uniform model universes of non-zero pressure when treated by general relativity.

A most important change has occurred in passing from eqn (5.9),

$$\dot{R}^2 = \frac{C}{R} + \frac{1}{3}\lambda R^2 - kc^2, \quad C \equiv \frac{8\pi}{3}G\rho R^3,$$

to eqn (9.12),

$$\dot{R}^2 = \frac{8\pi}{3}G\rho R^2 + \frac{1}{3}\lambda R^2 - kc^2.$$

It is that ρR^3 can no longer be assumed as constant in time: *this holds only for zero-pressure models* (see eqn (9.8)). Thus the models to be investigated now will exhibit dependences of ρ on the scale factor which have not been encountered in earlier sections.

It is convenient to regard eqn (9.12) together with eqn (9.13) as the first of our main equations for uniform universes with pressure. The second equation must then specify the pressure, and one could use eqn (9.1) with $\mu = 1$. However, it is more usual to use the equation derived by adding $(2/R)$ times eqn (9.1) to R^{-2} times eqn (5.9). This leads to cancellation of the term $(8\pi/3)G\rho$ and one finds

$$- \frac{8\pi G}{c^2} p = 2\frac{\ddot{R}}{R} + \left(\frac{\dot{R}}{R}\right)^2 + \frac{kc^2}{R^2} - \lambda. \qquad (9.14)$$

Thus eqns (9.12) and (9.14) can serve as our main equations. They are precisely the appropriate equations which are furnished by general relativity. In that sense there is no difference between the amended Newtonian theory and general relativity.

Alternatively one can use eqns (9.12) and (9.5) as the equation for the pressure. This will in fact be our procedure here. In Section 9.4 we shall use the first law of thermodynamics in the form (9.5). In Section 9.5 we shall apply the energy eqn (9.12) (or (5.9)).

9.3. THE THERMODYNAMICS OF SOME IDEAL FLUIDS

One may suppose that the cosmological fluid behaves locally like a simple fluid whose pressure is determined by an *equation of state*. This is a relation connecting three variables of a fluid. For example, *thermal equations of state* have the form

$$p = p(V,T),$$

and the *ideal classical gas* furnishes the best-known example in

$$p = NkT/V. \qquad (9.15)$$

Here T denotes the absolute temperature measured in degrees Kelvin, N is the number of molecules in volume V, and Boltzmann's constant is

$$k = 1.3806 \times 10^{-16} \text{ erg per degree Kelvin}$$
$$= 1.3804 \times 10^{-23} \text{ Joule per degree Kelvin } [JK^{-1}].$$

Here we shall use also the equation of state of an *ideal quantum gas*

$$pV = gU \quad \text{or} \quad p = g\rho c^2 \qquad (9.16)$$

where g is a constant and U is the internal energy.

It is shown in books on thermodynamics (see Problem 9.5) that

$$p = T\left(\frac{\partial p}{\partial T}\right)_V - \left(\frac{\partial U}{\partial V}\right)_T. \tag{9.17}$$

Multiplying by V/g and using eqn (9.16),

$$U = \frac{TV}{g}\left(\frac{\partial p}{\partial T}\right)_V - \frac{V}{g}\left(\frac{\partial U}{\partial V}\right)_T = T\left(\frac{\partial U}{\partial T}\right)_V - \frac{V}{g}\left(\frac{\partial U}{\partial V}\right)_T. \tag{9.18}$$

Suppose that there exists a function f such that

$$UV^g = f(TV^g) \equiv f(z), \quad z \equiv TV^g. \tag{9.19}$$

It follows from this assumption that

$$\left(\frac{\partial U}{\partial T}\right)_V = V^{-g}\frac{\mathrm{d}f}{\mathrm{d}z}\left(\frac{\partial z}{\partial T}\right)_V = \frac{\mathrm{d}f}{\mathrm{d}z}$$

and

$$\left(\frac{\partial U}{\partial V}\right)_T = -\frac{g}{V}U + \frac{gT}{V}\frac{\mathrm{d}f}{\mathrm{d}z}.$$

It is therefore seen that eqn (9.19) satisfies eqn (9.18) for U. In fact it can be shown that eqn (9.19) is the only solution (see Problem 9.6).

The second law of thermodynamics states that for *quasi-static processes* there exists a function S of state, called the *entropy* of the system, such that

$$\mathrm{d}S = T^{-1}(\mathrm{d}U + p\,\mathrm{d}V). \tag{9.20}$$

We can find properties of the entropy of an ideal quantum gas by multiplying eqn (9.20) by TV^g to find

$$TV^g\,\mathrm{d}S = V^g\left(\mathrm{d}U + \frac{gU}{V}\mathrm{d}V\right) = \mathrm{d}(UV^g).$$

It follows that

$$dS = z^{-1} df(z),$$ (9.21)

so that

$$\left.\begin{array}{l} \text{constancy of } UV^g, \text{ constancy of } z, \text{ constancy of } S \\ \text{are equivalent for an ideal quantum gas.} \end{array}\right\}$$ (9.22)

This result will be useful in Section 9.4 and in Problem 9.10.

We next wish to discuss the ideal classical gas. Observe that eqn (9.15) in eqn (9.17) yields

$$p = T\frac{Nk}{V} - \left(\frac{\partial U}{\partial V}\right)_T,$$

whence

$$(\partial U/\partial V)_T = 0 \qquad \text{(ideal classical gas)}$$ (9.23)

It follows that U is a function of T only. Hence the heat capacity at constant volume is

$$C_V = \left(\frac{\partial Q}{\partial T}\right)_V = T\left(\frac{\partial S}{\partial T}\right)_V = \left(\frac{dU}{dT}\right)_V = \left(\frac{dU}{dT}\right)$$ (9.24)

where eqn (9.20) has been used. The heat capacity at constant pressure, on the other hand, contains *two* terms:

$$C_p = T\left(\frac{\partial S}{\partial T}\right)_p = \left(\frac{\partial U}{\partial T}\right)_p + p\left(\frac{\partial V}{\partial T}\right)_p = \left(\frac{\partial U}{\partial T}\right)_p + Nk = \frac{dU}{dT} + Nk.$$ (9.25)

It follows that

$$C_p - C_V = Nk.$$ (9.26)

For a generalization of this result, see Problem 9.7.

It is easily shown that an ideal classical gas of constant C_p is in fact an ideal quantum gas. For C_V is then also a constant, so that eqn (9.24) yields

$$U = C_V T$$ (9.27)

if one assumes that $U \to 0$ as $T \to 0$. Multiplying eqn (9.27) by $Nk/(C_p - C_V)$,

$$U = \frac{C_V}{C_p - C_V} NkT = \frac{1}{\gamma - 1} pV = \frac{NkT}{\gamma - 1}, \quad \gamma \equiv C_p/C_v, \qquad (9.28)$$

where eqn (9.15) has been used. Since C_V, C_p are constants, γ is also a constant. Hence a form of eqn (9.16) has been obtained with

$$g = \gamma - 1 \quad \text{(classical ideal gas of constant heat capacities).} \qquad (9.29)$$

It can be shown from kinetic theory or from statistical mechanics that if the particles of the gas consist of structureless, weakly interactingmolecules moving in 1, 2, or 3 dimensions, the heat capacity is

$$C_V = \frac{1}{2}Nk, \quad Nk, \quad \text{or} \quad \frac{3}{2}Nk$$

respectively. In these cases eqn (9.26) shows that

$$\gamma = 3, \quad 2, \quad \frac{5}{3}$$

respectively. If the particles of the gas are diatomic molecules in three dimensions with two rotational degrees of freedom[†] a further term Nk is added to C_V so that

$$C_V = \frac{5}{2}Nk, \quad C_p = \frac{7}{2}Nk, \quad \gamma = \frac{7}{5} .$$

More generally, for particles with f degrees of freedom,

$$C_V = \frac{f}{2}Nk, \quad C_p = \frac{f+2}{2}Nk, \quad \gamma = 1 + \frac{2}{f} .$$

A particularly important type of ideal quantum gas is

[†] These are in this case rotations about the two axes perpendicular to the axis joining the atoms.

black-body radiation at temperature T. Such a system is specified by a special form of eqn (9.19), namely

$$g = \frac{1}{3}, \quad f(x) = ax^4, \quad a \equiv \frac{8}{15} \frac{\pi^5 k^4}{h^3 c^3} = 7.5641 \times 10^{-15} \text{ erg cm}^{-3} \text{ K}^{-4}$$
$$= 7.5641 \times 10^{-16} \text{J m}^{-3} \text{ K}^{-4}. \tag{9.30}$$

It follows from eqns (9.19) and (9.16) that

$$U = V^{-1/3} aT^4 V^{4/3} = aT^4 V = 3pV \tag{9.31}$$

and from eqn (9.21) that $dS = 4az^2 dz$, so that

$$TS = \frac{4}{3}az^3 T = \frac{4}{3}aT^4 V = \frac{4}{3}U . \tag{9.32}$$

The pressure can in this case be visualized by imagining radiation to be composed of particle-like entities, called photons, which would be reflected by a surface held in the system. They would in fact bounce off like ordinary particles. However, the energy of these particles arises from their motion and not from any inertial (rest) mass associated with them. Thus p stands in this case for radiation pressure. One may also associate a number of photons

$$\begin{aligned} N_r &= 60.42 \left(\frac{kT}{ch}\right)^3 V, \quad (16\pi\zeta(3) = 60.42), \\ &= \frac{30\zeta(3)}{\pi^4} \frac{a}{k} T^3 V = 0.3702 \frac{aT^3}{k} V \end{aligned} \right\} \tag{9.33}$$

with the radiation, where

$$\zeta(n) \equiv \sum_{j=1}^{\infty} j^{-n} \quad (n \geqslant 0), \quad \zeta(3) = 1.202 \tag{9.34}$$

is the so-called Riemann zeta function. Because energy is equivalent to mass, one can express the energy as equivalent mass density ρ_r per unit volume

$$\frac{U}{V} = aT^4 = \rho_r c^2 . \tag{9.35}$$

The entropy per unit volume is then

$$\frac{S}{V} = \frac{4}{3} \frac{\rho_r c^2}{T} .$$ (9.36)

9.4. SCALE FACTOR DEPENDENCES IN UNIVERSES WITH MATTER AND RADIATION

The early universe can probably be modelled by thinking of it as containing both matter and radiation, to be denoted by suffices m and r respectively. If $\rho(t)$ is the equivalent mass density and p the pressure, one then has

$$\rho = \rho_m + \rho_r, \quad p = p_m + p_r .$$ (9.37)

The matter content will be modelled by an ideal classical gas of constant heat capacities as discussed in Section 9.3. Let n_m be the number density of gas particles of rest mass m_0. The internal heat energy per unit volume at temperature T_m is then by eqn (9.28)

$$\frac{U_m}{V} = (\gamma-1)^{-1} n_m k T_m, \quad n_m \equiv \frac{N_m}{V}$$ (9.38)

where n_m is the number of gas particles per unit volume. To obtain ρ_m this has to be divided by c^2, and the mass density due to the rest mass of the particles has to be added:

$$\rho_m = n_m m_0 + (\gamma-1)^{-1} c^{-2} n_m k T_m .$$ (9.39)

The pressure contribution is obtained from eqn (9.15),

$$p_m = n_m k T_m .$$ (9.40)

The radiation is modelled by black-body radiation in the same volume, so that by eqns (9.3) and (9.31),

$$\rho_r = ac^{-2} T_r^4, \quad p_r = g\rho_r c^2 = \tfrac{1}{3} a T_r^4 .$$ (9.41)

We shall substitute eqns (9.39) to (9.41) into eqn (9.5) in the form

$$d(\rho R^3) + pc^{-2} dR^3 = 0$$

or equivalently

$$\frac{d}{dR}(\rho R^3) + 3pc^{-2}R^2 = 0.$$ (9.42)

This yields

$$\frac{d}{dR}\left[n_m m_0 R^3 + \frac{kT_m}{c^2(\gamma-1)}n_m R^3 + \frac{a}{c^2}T_r^4 R^3\right] + \frac{3R^2}{c^2}\left[n_m kT_m + \frac{1}{3}aT_r^4\right] = 0.$$ (9.43)

There are now a number of cases of interest.

(i) *Dust or incoherent matter alone*

These terms are applied to matter which does not exert any pressure. It follows directly from eqn (9.42) that $n_m R^3$ is independent of time. Thus, if C_m is independent of time, and using eqn (9.27),

$$\rho_m = C_m R^{-3}.$$ (9.44)

(ii) *An ideal quantum gas alone*

We shall consider the extreme relativistic region in which the energy due to the rest mass of the particles is neglected. Problem 9.10 deals with the appropriate generalization. Using eqn (9.16) in eqn (9.42),

$$R^3\frac{d\rho}{dR} + 3\rho R^2 + \frac{3}{c^2}g\rho c^2 R^2 = 0$$

so that

$$\frac{d\rho}{\rho} = -3(1+g)\frac{dR}{R}.$$

Thus an ideal quantum gas, when taken as the cosmological fluid, implies an expansion of the universe such that the following quantities are independent of time:

$$\rho R^{3(1+g)}, \quad \rho V^{1+g} = (\rho V)V^g, \quad UV^g, \quad S.$$ (9.45)

In the last step eqn (9.22) has been used. The expansion of dust is energy-conserving by eqn (9.44), but the expansion of an ideal quantum gas is entropy conserving, while its energy decreases during the expansion as R^{-3g}. This drop in energy is due to the work done by the pressure of the gas in the expansion. Apart from the constancy of S, case (i) is recovered if $g = 0$.

In particular for an ideal classical gas of constant heat capacities and constant particle numbers, eqn (9.43) yields $T_m R^{3(\gamma-1)} = B$, where B is independent of $R(t)$, i.e.

$$\left. \begin{aligned} T_m &= BR^{3(1-\gamma)} \quad [= BR^{-2} \text{ if } \gamma = \tfrac{5}{3}] \\ \rho_m &= \frac{m_0(n_m R^3)}{R^3} + \frac{Bk(n_m R^3)}{c^2(\gamma-1)}\frac{1}{R^{3\gamma}} \end{aligned} \right\} \tag{9.46}$$

For black-body radiation, on the other hand, eqn (9.45) with $g = 1/3$, together with eqn (9.31), shows that

$$\rho_r \propto R^{-4}, \quad T_r \propto \rho_r^{1/4} \propto R^{-1}. \tag{9.47}$$

(iii) *Non-interacting matter and radiation*

In this case eqn (9.42) splits into two equations, one for matter and one for radiation, and one finds eqns (9.46) and (9.47), so that

$$\rho = \frac{C_m}{R^\alpha} + \frac{C_r}{R^4}, \quad T_m \propto R^{3(1-\gamma)}, \quad T_r \propto R^{-1} \tag{9.48}$$

where α has the value 3 for dust and $3\gamma \sim 5$ when the gas pressure is not negligible. The gas and the radiation develop independently in the universe during the period in which the assumption of negligible interaction is applicable.

A rough idea of the possible circumstances in an expanding universe of matter plus radiation may be obtained as follows. Starting with a hot big bang, radiation and matter are initially in equilibrium. As the expansion proceeds, the radiation energy density drops more rapidly (as R^{-4}) than does the matter energy density ($\propto R^{-3}$). On the other hand, the situation is reversed for the temperature, the matter temperature dropping more rapidly ($\propto R^{-2}$) than the radiation temperature ($\propto R^{-1}$), as the two components cease to be in equilibrium.

This is, however, only a rough picture since matter obeys the thermodynamic laws of radiation at high compression when it is relativistic. Furthermore the creation of heavier nucleons as expansion proceeds has to be taken into account (see Section 1.4). However, these simple ideas can also be extended to oscillating models of the universe, but they have to be corrected for the interaction between matter and radiation [3].

(iv) *Matter and radiation in thermal equilibrium*
     The simplest interaction between matter and radiation which one can suppose is that these two systems are in thermal equilibrium: $T_m = T_r$. If it is also assumed that the number of particles in the gas is time independent, then eqn (9.43) yields, after multiplying by $c^2/3n_m k$,

$$\left(\frac{1}{3(\gamma-1)}+s\right)R^3 \frac{dT}{dR} + sTR^2 + TR^2 = 0,$$

where

$$s \equiv \frac{4aT_r^3}{3n_m k} \tag{9.49}$$

is the photon entropy eqn (9.32) per particle of the gas in units of k. It follows that

$$\frac{dT}{T} = -\frac{1+s}{[3(\gamma-1)]^{-1}+s} \frac{dR}{R}. \tag{9.50}$$

Two limiting cases are of interest:
     (a) $0 < s \ll 1$. In this case

$$\left. \begin{array}{l} TR^{3(\gamma-1)} \text{ is independent of time as for} \\ \text{the ideal classical gas.} \end{array} \right\} \tag{9.51}$$

     (b) $s \gg 1$. In this case

$$TR \text{ is independent of time as for radiation.} \tag{9.52}$$

In the first case one would expect s to behave as

$$s \propto \frac{T^3}{n_m} \propto \frac{R^{-9(\gamma-1)}}{R^{-3}} = R^{12-9\gamma} \left[= R^{-3} \text{ for } \gamma = \frac{5}{3} \right]. \quad (9.53)$$

Thus if s was small in the past, it will remain so. In the second case one would expect

$$s \propto \frac{T^3}{n_m} \propto \frac{R^{-3}}{R^{-3}} . \quad (9.54)$$

So if s was once large it will remain large, since it is roughly constant in time. This second case appears to be the one which is realized (see relation (9.60)).

9.5. SOME NUMERICAL VALUES

The number density of photons in black-body radiation at temperature T is given by eqn (9.33):

$$n_r = \frac{8\pi^5}{45} \frac{\zeta(3)}{\pi^4/90} \left(\frac{kT}{ch}\right)^3 = 60.42 \times \left(\frac{0.13806\times10^{-22}}{2.998\times10^{+8}\times0.6625\times10^{-33}}\right)^3 T^3$$

$$= 20.30 [T(K)]^3 \, 10^6 \, (m^{-3}). \quad (9.55)$$

One can re-express s of eqn (9.49) in terms of n_r as

$$s = \frac{2\pi^4}{45\zeta(3)} \frac{n_r}{n_m} = 3.6017 \frac{n_r}{n_m} . \quad (9.56)$$

If we take the present matter density as

$$5 \times 10^{-28} \text{ kg m}^{-3} \lesssim \rho_{m0} \lesssim 2 \times 10^{-26} \text{ kg m}^{-3} \quad (9.57)$$

then for protons of rest mass 1.67×10^{-27} kg this implies a material number density

$$0.3 \text{ m}^{-3} \lesssim n_{m0} \lesssim 11 \text{ m}^{-3}. \quad (9.58)$$

The main contribution to the present photon number density is the recently discovered background black-body radiation at temperature 2.7 K, which by eqn (9.55) yields

$$n_{r0} \sim 4 \times 10^8 \, (m^{-3}). \quad (9.59)$$

Hence eqn (9.56) yields

$$1.3 \times 10^8 \lesssim s_0 \lesssim 53 \times 10^8 \; , \tag{9.60}$$

which is a large number, so that eqns (9.50) and (9.54) hold.
Therefore if s has remained broadly constant in time, a large
value of s can be assumed in the treatment of the early universe. The present number of photons per proton is, by eqns
(9.58) and (9.59),

$$\eta_0 = \frac{n_{r0}}{n_{m0}} \sim \frac{4 \times 10^8}{1} \sim 4 \times 10^8.$$

9.6. TIME DEPENDENCES IN EARLY UNIVERSES WITH MATTER AND RADIATION

If the scale factor is small enough, as it has been for the
early or young universe, ρ is large enough to enable one to
simplify eqn (9.12) to

$$\dot{R}^2 = \frac{8\pi}{3} G\rho R^2 \; . \tag{9.61}$$

The high densities in the early universe led to a rapid estab-
lishment of thermal equilibrium between matter and radiation,
so that case (iv) of Section 9.4 is relevant. Also $s \gg 1$ from
Section 9.5, so the radiation regime of case (ii) is a good
approximation by eqn (9.50). Accordingly,

$$\rho \propto R^{-4} >, \text{ i.e. } \dot{\rho}/\rho = -4\dot{R}/R$$

can be used in eqn (9.61), whence

$$\frac{\dot{\rho}}{\rho} = -4\sqrt{\left(\frac{8\pi}{3}G\right)}\sqrt{(\rho)}$$

i.e.

$$\rho = \frac{3}{32\pi G t^2} \; . \tag{9.62}$$

This also implies a statement about the time dependence of the
temperature since

$$\rho c^2 = \vartheta a T^4$$

where the numerical factor ϑ depends on the precise content
of the universe. If black-body radiation dominates, $\vartheta \sim 1$.
But if relativistic particles contribute sensibly, then $\vartheta > 1$.
It follows that

$$\dot{T} = \left(\frac{3c^2}{32\pi\vartheta a G}\right)^{1/4} t^{-1/2}. \qquad (9.63)$$

The value of the coefficient is (with $G/c^2 = 7.425 \times 10^{-28}$ m
kg^{-1})

$$\left(\frac{3}{32\pi\vartheta \times 7.564\times10^{-16} \times 7.425\times10^{-28}}\right)^{1/4} \sim 1.52(\vartheta)^{-1/4}\ 10^{10},$$

so that the temperature in degrees Kelvin as a function of
time in seconds is

$$T = \frac{1.52\times10^{10}}{\vartheta^{1/4}\sqrt{(t)}}. \qquad (9.64)$$

Various corrections have to be applied to this result. However, its wide rough applicability is due to a characteristic
of quantum statistics. This is that systems of particles of
non-zero rest mass share many of their thermodynamic properties with black body radiation, provided only the temperature
is high enough, so that the particles have an average energy
which is far in excess of their rest energy.

    Fig. 9.1 gives an idea of the presumed thermal history of
the universe. The curve for the radiation temperature roughly
follows eqn (9.64). The densities of matter (baryons), and
of electrons are also shown. At 10^6 years or 3×10^{13} s,
matter and radiation 'decouple' and cool down separately
according to regime (iii), the matter temperature falling more
rapidly than the radiation temperatures in accordance with
eqns (9.46) and (9.47).

    We shall now use the property $p_r = \frac{1}{3}\rho_r c^2$ (eqns (9.16)
and (9.30)) of black-body radiation for all the contents of
the early universe. This is justified because a system of
particles moving at relativistic speeds does in fact satisfy

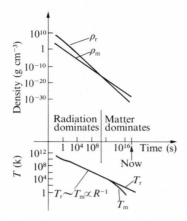

FIG. 9.1   The presumed thermal history of the universe [5].

this relation.   It is therefore applicable to the whole contents of the early universe to a good degree of approximation.
Eqn (9.62) then yields

$$p = \frac{c^2}{32\pi G t^2} .$$

For this simple case we have therefore obtained a rather complete solution of the cosmological problem by finding ρ, p, and
R as functions of the time.   Contrary to what one would expect,
the universe has developed its non-equilibrium characteristics
with lapse of time: T_m has dropped below T_r, and both ρ_m and
ρ_r decrease now more rapidly than either, as seen in eqn (9.48).
        Slightly more complicated models can be made which also
allow complete solutions to be obtained by analytical methods
[4].

9.7. SUMMARY OF THE DYNAMICAL EQUATIONS
For uniform model universes two energy equations have been
encountered.   One of these arose from the Newtonian model,
eqn (5.9), and was seen to hold also for relativistic universes of non-zero pressure, eqn (9.12).   This dealt with the
total energy of gravitating matter enclosed within a surface

S* which was assumed frozen into the matter. The energy was gravitational and kinetic:

$$\dot{R}^2 = \frac{8\pi G}{3}\rho R^2 + \frac{1}{3}\lambda R^2 - kc^2. \tag{9.65}$$

The other equation dealt with the effect of pressure at the surface, eqn (9.3) or (9.6) or (9.7):

$$1 + \frac{p}{c^2\rho} + \frac{1}{3}\frac{\dot{\rho}}{\rho}\frac{R}{\dot{R}} = 0. \tag{9.66}$$

It is convenient to introduce again the following abbreviations

$H \equiv \dot{R}/R =$ (Hubble's 'constant'), $q \equiv -R\ddot{R}/\dot{R}^2 =$
(the deceleration parameter)   (9.67)

$g \equiv p/c^2\rho$ (a constant for the ideal quantum gas, but not otherwise, see Problem 9.4. It is a dimensionless quantity).   (9.68)

$\sigma \equiv 4\pi G\rho/3H^2$ the dimensionless energy parameter, ρc^2 being the energy density   (9.69)

$L \equiv \lambda/3H^2$, $K \equiv kc^2/R^2H^2$, two more dimensionless quantities.   (9.70)

The energy eqn (9.12) is then

$K - L = 2\sigma - 1$ (energy equation)   (9.71)

Several of these results generalize those of Section 4.6.

Differentiating eqn (9.65) with respect to time and eliminating by (9.65), one recovers the equation of motion (9.1):

$$\frac{\ddot{R}}{c^2R} + \frac{4\pi G}{3c^2}\left(\rho + 3\frac{p}{c^2}\right) - \frac{\lambda}{3c^2} = 0. \tag{9.72}$$

Multiplying by c^2/H^2, this can be written as

$$q + L = (1+3g)\sigma \quad \text{(equation of motion).} \quad (9.73)$$

A number of other relations can be derived from these two. For example, subtracting R^{-2} times (9.65) from (9.72), one finds

$$4\pi G\left(\rho + \frac{p}{c^2}\right) = \frac{kc^2}{R^2} + \left(\frac{\dot{R}}{R}\right)^2 - \frac{\ddot{R}}{R}. \quad (9.74)$$

After division by H^2 this can be written as

$$q + K = 3(1+g)\sigma - 1. \quad (9.75)$$

This result is also obtainable by adding eqns (9.71) and (9.73).

An important equation, because it is one of the Einstein field equations for these models, is (9.14). This can be written as

$$2q - K - 3L = 1 + 6g\sigma. \quad (9.76)$$

This result is also obtainable by subtracting eqn (9.75) from three times (9.73).

For zero-pressure models ($g = 0$) one recovers eqn (4.24) and (4.27) in the forms below:

from eqn (9.71): $K - L = 2\sigma - 1$ (energy equation) $\quad (9.77)$

from eqn (9.73): $q + L = \sigma$ (equation of motion) $\quad (9.78)$

from eqn (9.75): $q + K = 3\sigma - 1$ (an implication). $\quad (9.79)$

In the general theory of relativity it is shown that eqn (9.65) or (9.12) and (9.14) are the Einstein field equations for these models, and the other equations obtained here occur as consequences. In this sense all equations obtained are relativistic equations.

PROBLEMS

On Section 9.1

(9.1)   In some hybrid of the Newtonian theory and the amended
Newtonian theory one might allow for non-zero pressure in
the energy conservation eqn (9.3) (or (9.5), (9.6), or
(9.7)), thus using the amended Newtonian theory. But one
might put $\mu = 0$ in eqn (9.1), thus using the Newtonian
theory. Show that in this case

$$\dot{R}^2 = -\frac{8\pi G}{3} \int_{t_2}^{t} \rho R \ dR + \frac{1}{3}\lambda\dot{R} + \left\{\dot{R}^2 - \frac{1}{3}\lambda R^2\right\}_{t_2} . \tag{9.80}$$

(9.2)   Replace G by an effective gravitational 'constant' G_N
defined by the requirement that Newtonian and relativistic theory shall yield identical results. Hence show
that

$$\frac{G_N}{G} = \frac{(\rho R^2)_{t_2} - (\rho R^2)_t}{\int_{t_2}^{t} \rho R \ dR} .$$

(9.3)   Show from the result of Problem 9.2 that $G_N = G$ for
zero-pressure models.

(9.4)   An ideal quantum gas [6] is a fluid such that $p = gc^2\rho$
where g is a constant. Show that if the cosmic fluid is
such a fluid, then
   (i) $\rho(t)[R(t)]^{3+3g}$ is independent of time, and hence
   (ii) G_N is a constant whose value is $1 + 3g$ times the
         normal gravitational constant. Note that for $g = 0$
         one recovers a special case of the result of Prob-
         lem 9.3.

On Section 9.3

(9.5) (i)   The Gibbs free energy of a system is $G \equiv U + pV - TS$.
             Show by considering dG that $\left(\frac{\partial p}{\partial T}\right)_V = \left(\frac{\partial S}{\partial V}\right)_T$.

(ii)   From the second law, using the result of (i), establish eqn (9.17).

(9.6)   Show that eqn (9.19) is the only solution of (9.18).

(9.7)   Show that for any simple fluid, eqns (9.24) and (9.25) can be generalized to $C_p - C_V = \left[\left(\frac{\partial U}{\partial V}\right)_T + p\right]\left(\frac{\partial V}{\partial T}\right)_p$.

[Hint: $\left(\frac{\partial U}{\partial T}\right)_p - \left(\frac{\partial U}{\partial T}\right)_V = \left(\frac{\partial U}{\partial V}\right)_T\left(\frac{\partial V}{\partial T}\right)_p$ may be established first.]

(9.8)   An ideal quantum gas which satisfies the additional relation $TS = jU$ where j ($\neq 1$) is a constant is called *superideal*. Show that the function $f(z)$, $z = Tv^g$, has then the form of the simple power law $f(z) \propto z^{j/(j-1)}$. Apply this result to black-body radiation [7].

On Section 9.4

(9.9)   A classical ideal gas has a constant heat capacity ratio γ and contains a conserved number $N_m = n_m V$ of molecules of rest mass m_0. Show that if this system serves as the cosmological fluid then the mass density as a function of scale factor has the form

$$\rho_m = \frac{m_0(n_m R^3)}{R^3} + \frac{C}{R^{3\gamma}}$$

where the constant C is given by

$$\frac{C}{R^{3\gamma}} = \frac{C_V}{c^2 V} T_m.$$

Here V is the volume of the model universe and T_m is the temperature.

(9.10)   Generalization of case (ii) eqn (9.45). An ideal quantum gas of particles whose rest mass is m_0 satisfies

$$pV = gU = g(\rho V - Nm_0)c^2.$$

If the number of particles $N = nV$ is constant, show that
for a cosmological fluid consisting of such a system

$$\rho = \frac{m_0 (nR^3)}{R^3} + \frac{C}{R^{3(g+1)}}$$

where C is a constant. Show also that the entropy of the
system is conserved during the expansion or contraction
of the model universe.

On Section 9.6

(9.11) Verify the table below for radiation temperature as a
        function of time

		←radiation dominates→			
	1 s	10 s	100 s	1000 s	1 week
Temperature (K)	1.52×10^{10}	3.16×10^{9}	1.52×10^{9}	3.16×10^{8}	1.29×10^{7}

			←matter dominates→
	30 days	1 year	10^6 years	10^{10} years (Now)
	$7.42 \ 10^6$	$1.78 \ 10^6$	1780	17.8

[Use eqn (9.63).]

(9.12) Find the values of ρ_{r0} and ρ_{m0}, the equivalent matter
        densities due to radiation and nucleons at the present
        time and verify that

$$\rho_{m0}/\rho_{r0} \sim 10^3.$$

[Assume the radiation is due to the background black-body
radiation at 2.7 K.]

On Section 9.6

(9.13)(i) Show on the same basis that at present the ratio of
          the number of photons to the number of nucleons is
          of order 10^9.

   (ii) If the nucleons were at present all in the form of
        hydrogen atoms, show also that if they were to be
        all ionized at the expense of energy in the back-
        ground radiation, the temperature of that radiation

would drop only by a fraction which is less than 10^{-5} of its present temperature of 2.7 K.

On Section 9.7

(9.14)   Prove that $qK - (1+q)L = (2q-3g-1)\sigma$.

(9.15)   Prove that $2q + 3(1+g)L - (1+3g)(K+1) = 0$.

(9.16)   Prove that $\sigma = C/2R^3H^2$, where C is given by eqn (5.9).

OPTICAL EFFECTS OF THE EXPANSION ACCORDING
TO VARIOUS THEORIES OF LIGHT

10.1. THE PROBLEM OF THE ETHER

The observations which have formed our present picture of the
universe, and those which may lead in the future to a decision
between cosmological models, almost all involve the reception
of light or other electromagnetic radiations (the word 'light'
will often be used where 'electromagnetic radiation' would be
more accurate). In order to discuss the observational consequences of any cosmological model it is therefore necessary to
supplement it with a model of light propagation. Classical
physics offers two obvious models for this purpose; the particle model, which was preferred by Newton, and the wave model.
In the particle model light emission is pictured as somewhat
analogous to the firing of cannon balls whose trajectories
correspond to light rays. One knows that frequencies have to
be associated with these particles of light, and classical
physics provides no simple way of doing this. One has to go
to the quantum theory in order to obtain the appropriate relations.

   A more convenient model for the purposes of this chapter
is therefore the classical wave theory of light, which can be
summarized as follows: light consists of waves, which travel
in interstellar space in straight lines with a constant speed
c relative to the ether. We shall also need the result of
quantum theory that a light wave, when it interacts with
matter, can be considered as a series of particles or photons,
and that the energy E of each photon is proportional to its
frequency ν: $E = h\nu$, where h is Planck's constant. For our
purposes the ether is needed merely as a reference frame, and
the question of its essential nature does not arise. However,
the expansion of the universe poses some awkward questions in
the wave theory; is the ether stationary, so that all galaxies
(or all but one) are moving through it, or does it take part
in the expansion? The first possibility would mean that most

of the galaxies in the universe are moving through the ether
at speeds comparable to that of light.   Since our own galaxy
is certainly not moving at such a speed (or the wavelength of
light would not serve as a precise standard of length) we
would be in a highly privileged position, and cosmologists
tend to distrust any such singling out of our own galaxy (see
Section 2.2).   The alternative, that the ether expands with
the galaxies, poses other problems — does the ether become
less dense with time, and does light (and all electromagnetic
interaction) therefore become weaker?   Fortunately these questions were never to impede the progress of cosmology, since by
the time the expansion was discovered, Einstein had already
provided, in his special and general theories of relativity,
wave models of light propagation which did not require an
ether.

The use of *classical optics* in Sections 10.2 to 10.5 is,
therefore, an attempt to understand the basic optical pheno-
mena in their simplest possible setting.   To do this we shall
assume that the ether does *not* take part in the expansion, and
also that astronomical observations are made by an observer at
rest in the ether.   This violates the cosmological principle,
but has the advantage that some important equations can then
be easily derived in the precise form which they take in the
relativistic models (Sections 10.6 and 10.7).   The main
results of all three models are collected in Tables 10.1 and
10.2, for use in discussing cosmological observations in
Chapter 11.   They serve as a summary of this chapter, apart
from the last section.

Although the general relativistic model is generally
believed to be the most accurate for observations of objects
at cosmological distances, or with intense gravitational
fields, the other models have their own spheres of usefulness.
The classical model is valid for observations of nearby, slow-
moving objects (velocity $v \ll c$ and distance $r \ll ct_H$) and offers
extreme mathematical simplicity.   It is used for the Andromeda galaxy and other nearby galaxies.   The special relativistic model is valid for observations of nearby fast-moving
objects ($v \sim c$ but $r \ll ct_H$) which may include some quasars [1],

TABLE 10.1

Observational quantities as functions of red-shift z
for the classical wave theory, special, and general relativity

---

Factors by which the apparent brightness
of a receiving source is multiplied:

Red-shift factor, $f_{rs} = (1+z)^{-2}$ (eqn 10.7)

Aberration factor, $f_{a} = (1+z)^{-2}$ (eqn 10.15)

Frequency response factor $f_{\nu} = (1+z)^{1-s}$ (eqn 10.13)
for a radio source with
spectral index s

Ratio of apparent to real
density of sources $N_{r}/N_{rE} = 1 + z$ (eqn 10.28)

Bolometric luminosity
distance of a source
observed at distance r, $d_{L} = (1+z)^{2}r$ (eqns 10.21 and 10.47)

TABLE 10.2

Equations relating the red-shift z of a source to its speed of
recession v and to the time of emission t_E of the light by which
it is observed for various theories of light propagation

	Theory of light propagation		
	Classical	Special relativity	General relativity
Frequency ratio $\nu_E/\nu_0 = 1 + z$	$1 + v/c$	$\left\{ \dfrac{1+v/c}{1-v/c} \right\}^{\frac{1}{2}}$	$\dfrac{R(t_0)}{R(t_E)}$
	(eqn 10.5)	(eqn 10.29)	(eqn 10.43)
Speed of recession = v	cz	$\dfrac{cz(2+z)}{(1+z)^{2}+1}$	–
Scale factor at time of emission = $R(t_E)$	–	–	$R(t_0)/(1+z)$

The last two lines of the table contain simple algebraic consequences of
the first line.

and its use in that context is illustrated in Problem A.3.

10.2. THE RED-SHIFT AND ALLIED EFFECTS

The name 'red-shift' indicates that visible light emitted from a receding source with frequency ν_E appears to the observer to have a frequency ν_0 which is less than ν_E, so that the light appears to have been shifted towards the low-frequency or red end of the visible spectrum. The amount of the red-shift, z, is defined by

$$z \equiv \frac{\nu_E - \nu_0}{\nu_0}; \quad \nu_0 = \frac{\nu_E}{1+z} \qquad (10.1)$$

As a rule suffices 0 and E will refer respectively to the time and place at which a photon is observed, and the time and place at which it is emitted.

The calculation of the red-shift can be understood by reference to Fig. 10.1. Let two successive wavecrests of light leave the galaxy G, which is receding with speed v, at times t_E and $t_E + \Delta t_E$, and let them arrive at the observer O at times t_0 and $t_0 + \Delta t_0$. Thus the emitted frequency ν_E is $1/\Delta t_E$ and the observed frequency ν_0 is $1/\Delta t_0$, so that from eqn (10.1)

$$1 + z = \frac{\nu_E}{\nu_0} = \frac{\Delta t_0}{\Delta t_E} . \qquad (10.2)$$

Now suppose that the distance OG is equal to r at time t_E and to $r + \Delta r$ at time $t_E + \Delta t_E$. Then the first wavecrest travels the distance r in time $(t_0 - t_E)$ and the second travels

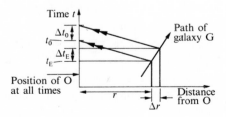

FIG. 10.1 Space—time diagram for the calculation of the red-shift in O's reference frame.

the distance $r + \Delta r$ in time $t_0 + \Delta t_0 - (t_E + \Delta t_E)$. Since both
wavecrests travel with speed c relative to O it follows that

$$t_0 - t_E = \frac{r}{c}, \quad t_0 - t_E + \Delta t_0 - \Delta t_E = r + \frac{\Delta r}{c}$$

or, subtracting the first equation from the second and re-
arranging,

$$\Delta t_0 = \Delta t_E + \frac{\Delta r}{c} \qquad (10.3)$$

Also, the galaxy moving with speed v must cover a distance
$\Delta r = v \Delta t_E$ in time Δt_E.   Substituting this into (10.3) gives

$$\Delta t_0 = \Delta t_E \left\{ 1 + \frac{v}{c} \right\} \qquad (10.4)$$

which with (10.2) gives the required expression for z:

$$1 + z = \frac{\Delta t_0}{\Delta t_E} = 1 + \frac{v}{c}, \quad \text{i.e. } z = \frac{v}{c}. \qquad (10.5)$$

For a source approaching with speed v one would have
$\Delta r = -v \Delta t_E$ and so $z = -v/c$.

As well as its obvious effect of shifting spectral lines
and so giving a measure of the speed of recession, the red-
shift has effects on the apparent brightness of the source.
In this context *apparent brightness* is defined as the amount
of energy radiated from the source, falling on a unit area
near O in unit time and within the operating frequency range
of a particular instrument. The instrument may be a human
eye, a radio telescope or a camera/telescope combination,
sensitive in each case to a fairly narrow frequency range;
alternatively it may accept radiated energy of all possible
frequencies. One can now envisage three ways in which the
motion of a receding source may alter its apparent brightness:

   (i)   the fraction of emitted photons which will ultimately
         reach the unit area in question may be altered;
   (ii) considering only those photons which will reach the
         unit area, the number arriving per unit time, and the
         energy of each individual photon, may be altered; and

(iii)   the frequency shift may cause the instrument to respond
        to a different range of emitted frequencies.

Of these effects (i) is not directly linked to the red-
shift, and is discussed in Section 10.3, but (ii) and (iii)
depend directly on the red-shift. Suppose that n_E is the number of photons emitted per unit time in such a direction as to
reach a certain instrument, and E_E is the mean energy of these
photons, both measured in the reference frame of the source G,
and that n_0, E_0 are the number arriving per unit time and the
mean energy on arrival, both measured in the reference frame
of the observer O. Then the effect (ii) of the red-shift is
to multiply the rate of energy reception, i.e. the apparent
brightness, by a factor

$$f_{rs} \equiv \frac{n_0 E_0}{n_E E_E} \tag{10.6}$$

compared to a stationary source. Now a figure exactly analogous to Fig. 10.1 could be drawn in which Δt_E, Δt_0 are the
time intervals between emission or arrival of successive photons, rather than wavecrests. One would then have
$n_E = 1/\Delta t_E$, $n_0 = 1/\Delta t_0$ and so

$$\frac{n_0}{n_E} = \frac{\Delta t_E}{\Delta t_0} = \frac{1}{1+z} \ .$$

Also, the quantum-mechanical relationship $E = h\nu$ is assumed to
hold in both reference frames, so that

$$\frac{E_0}{E_E} = \frac{h\nu_0}{h\nu_E} = \frac{\nu_0}{\nu_E} = \frac{1}{1+z} \ .$$

Combining these results gives finally

$$f_{rs} = \frac{1}{(1+z)^2} \tag{10.7}$$

as the main factor by which the red-shift multiplies the
apparent brightness of a receding source. This factor is
present for all types of instrument.

When the instrument has a limited frequency range the
apparent brightness is multiplied by a further factor, which

will be called f_ν, resulting from the effect (iii) of the red-
shift. Suppose that the frequency spectrum of the source can
be described by a function $P(\nu_E)$ in such a way that the rate
of energy output of the source in an emitted frequency range
$(\nu_1\nu_2)$ is

$$E(\nu_1\nu_2) \equiv \int_{\nu_1}^{\nu_2} P(\nu_E)\ d\nu_E = \int_{\nu_1}^{\nu_2} P(x)\ dx \qquad (10.8)$$

where the change from ν_E to the dummy variable x has been made
for later convenience. Suppose also that the instrument in
use registers all photons with observed frequencies between
ν_{10} and ν_{20}, and no others. Then the apparent brightness of
the source, if it were stationary (in the ether), would be
proportional to $E(\nu_{10},\nu_{20})$ as defined in eqn (10.8). However,
the motion has the effect that photons with *observed* frequencies between ν_{10} and ν_{20} are those with *emitted* frequencies
between ν_{1E} and ν_{2E}, where

$$\nu_{1E} = (1+z)\nu_{10}, \quad \nu_{2E} = (1+z)\nu_{20} \qquad (10.9)$$

It follows that the apparent brightness of the source is in
fact proportional to

$$E(\nu_{1E},\nu_{2E}) \equiv \int_{\nu_{1E}}^{\nu_{2E}} P(x)\ dx \qquad (10.10)$$

and the appropriate correction factor f_ν is therefore

$$f_\nu \equiv \frac{E(\nu_{1E},\nu_{2E})}{E(\nu_{10},\nu_{20})} \qquad (10.11)$$

The effect is illustrated for a hypothetical spectrum $P(\nu_E)$ in
Fig. 10.2. Evidently f_ν can be greater or smaller than 1
depending on the shape of the spectrum, and will not in general
be a simple function of z.

For observations in the visible region of the spectrum the
calculation of f_ν is further complicated by the fact that the
instrument, which usually incorporates a photographic film,

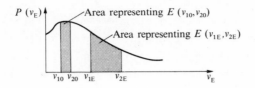

FIG. 10.2 Effect of the red-shift in causing an instrument to register a different part of the spectrum of a source. The figure is drawn for $z = 2$ in (10.9).

does not respond equally to all frequencies within its range. In the radio region things are sometimes simpler, because it often happens that

(a)   the response of the instrument is uniform over the frequency range (ν_1, ν_2)

(b)   the spectrum of the source can be represented over the range of interest by a power law

$$P(\nu_E) = K\nu_E^{-s} \qquad (10.12)$$

where K and s are independent of ν_E and s is known as the *spectral index* of the source. Because of (a) the correction factor f_ν is given accurately by eqn (10.11). Substituting eqn (10.12) into eqns (10.8) and (10.10), one finds

$$E(\nu_1, \nu_2) = \frac{K}{1-s}\left\{\nu_2^{1-s} - \nu_1^{1-s}\right\},$$

$$(s \neq 1)$$

$$E(\nu_{1E}, \nu_{2E}) = \frac{K}{1-s}\left\{\nu_{2E}^{1-s} - \nu_{1E}^{1-s}\right\}.$$

Substituting for ν_{1E} and ν_{2E} from eqn (10.9) and using the result in eqn (10.11) gives finally

$$f_\nu = (1+z)^{1-s} \quad \text{(radio source with spectral index } s) \quad (10.13)$$

Taking the value $s = 0.7$ as typical, we find $f_\nu = (1+z)^{0.3}$, a rather small correction compared to f_{rs}. Eqn (10.13) is also valid for $s = 1$, although a separate discussion is needed to establish this (see Problem 10.1)).

The calculation of the factors f_{rs} and f_v given here merely assumes (a) the existence of the red-shift factor $(1+z)$, and (b) that the same factor reduces the arrival rate of photons relative to their emission rate. No particular theory of light propagation is involved. Therefore, f_{rs} and f_v are the same functions of z in the relativistic theories as in the classical theory considered so far. However, the relation between z and the speed of recession v is quite different in those theories, as explained in Sections 10.6 and 10.7.

10.3. THE ABERRATION OF LIGHT

The name aberration is given to any change in the apparent direction of a light ray consequent on the motion of the source or observer. In the present context we consider that the source G is moving, as illustrated in Fig. 10.3.

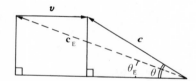

FIG. 10.3 Velocity addition diagram for the aberration of light, relating the velocity **c** of a light ray relative to O to its velocity $\mathbf{c}_E$ relative to G.

Here G emits a light ray which an observer moving with G considers to move with velocity $\mathbf{c}_E$ at an angle ϑ_E to the line GO. For O the velocity of the light ray is **c**, composed of the light ray's velocity $\mathbf{c}_E$ relative to G and G's own velocity **v** relative to O: $\mathbf{c} = \mathbf{c}_E + \mathbf{v}$. The geometry of the two right-angled triangles in the figure shows that

$$c_E \sin \vartheta_E = c \sin \vartheta$$

$$c_E \cos \vartheta_E = c \cos \vartheta + v$$

or dividing

$$\tan \vartheta_E = \frac{\sin \vartheta}{\cos \vartheta + v/c} \qquad (10.14)$$

The effect of aberration on the amount of light received from a source at distance r can be seen by reference to Fig. 10.4. This shows (solid lines) the paths of a set of light rays which all make the same angle ϑ with the line GO in O's reference frame, and (dotted lines) the paths which the same rays would take if the source were at rest. In O's reference frame the rays define a circle of radius $r\vartheta$ and area $\pi(r\vartheta)^2$ centred on O. If the source were at rest the same rays would define a smaller circle of area $\pi(r\vartheta_E)^2$. Thus the light which would have covered an area $\pi(r\vartheta_E)^2$ if the source were at rest is spread out by aberration to cover a larger area $\pi(r\vartheta)^2$, leading to a correction factor by which the energy received per unit area per unit time must be multiplied to allow for aberration:

$$f_a = (\vartheta_E/\vartheta)^2.$$

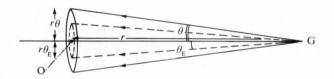

FIG. 10.4 The spreading out of light rays by aberration.
————➤——— direction of light rays relative to O; — — ➤ — — direction of light rays relative to G.

Using eqns (10.14) and (10.5) with the approximations (valid for small angles) $\tan \vartheta_E \doteqdot \vartheta_E$, $\sin \vartheta \doteqdot \vartheta$, $\cos \vartheta \doteqdot 1$ gives

$$f_a = \frac{1}{(1+v/c)^2} = \frac{1}{(1+z)^2} \qquad (10.15)$$

Thus aberration reduces the energy received from a receding source.

10.4. THE APPARENT BRIGHTNESS OF A SOURCE

Consider a source of electromagnetic radiation which is observed at a distance r, and define the following four quantities:

> L is the total energy output of the source per unit time,
> measured in the reference frame of the source,
> $A \equiv L/4\pi$ is the corresponding energy output per unit
> solid angle,
> S is the rate at which energy from the source falls on
> unit area at the observer O in unit time,
> d_L is defined by

$$d_L \equiv (A/S)^{\frac{1}{2}}. \tag{10.16}$$

The last three of these quantities have individual names:
A is the (bolometric) intrinsic brightness of the source,
S is the (bolometric) apparent brightness of the source,
d_L is the (bolometric) luminosity distance of the source.
The term 'bolometric', which will often be omitted for convenience, means that one is considering the entire spectrum of
radiation from a particular source. Corresponding (non-
bolometric) quantities which relate only to a restricted frequency range are given the suffix ν. Thus, for instance, the
symbol S_ν will always denote the rate at which energy from
the source falls on unit area in unit time within the frequency range of a particular instrument. S_ν may be an optical,
photographic, or radio apparent brightness depending on the
instrument considered. The corresponding luminosity distance
is defined by

$$d_{L\nu} \equiv (A_\nu/S_\nu)^{\frac{1}{2}}. \tag{10.17}$$

The reason for the name 'luminosity distance' can be seen
as follows. If the source were stationary, its light output
L would be spread uniformly over a sphere of area $4\pi r^2$ when it
reached O, and therefore

$$S \text{ (stationary source)} = \frac{L}{4\pi r^2} = \frac{A}{r^2}. \tag{10.18}$$

The same conclusion holds for a restricted frequency range, since there is no red-shift, and so

$$S_\nu \text{ (stationary source)} = \frac{A_\nu}{r^2} . \qquad (10.19)$$

It follows that, *for a stationary source*,

$$r = \left(\frac{A}{S}\right)^{\frac{1}{2}} = \left(\frac{A_\nu}{S_\nu}\right)^{\frac{1}{2}} ,$$

so that an observer who believes the source to be stationary will infer that its distance is $(A/S)^{\frac{1}{2}} \equiv d_L$ or $(A_\nu/S_\nu)^{\frac{1}{2}} \equiv d_{L\nu}$, depending on the type of observation made.

   If the source is not stationary, its bolometric apparent brightness must be multiplied by the factors f_{rs} and f_a of eqns (10.7) and (10.15) so that eqn (10.18) is replaced by

$$S = f_{rs}f_a \frac{A}{r^2} = \frac{A}{(1+z)^4 r^2}, \qquad (10.20)$$

and the luminosity distance (10.16) becomes

$$d_L = (f_{rs}f_a)^{-\frac{1}{2}}r = (1+z)^2 r. \qquad (10.21)$$

   A non-bolometric apparent brightness will also be multiplied by the factor f_ν of eqn (10.11), so that in general

$$S_\nu = f_{rs}f_a f_\nu \frac{A_\nu}{r^2} = \frac{A_\nu f_\nu}{(1+z)^4 r^2} \qquad (10.22)$$

and

$$d_{L\nu} = (1+z)^2 (f_\nu)^{-\frac{1}{2}}r. \qquad (10.23)$$

This result simplifies somewhat for a radio apparent brightness, when f_ν may be of the form (10.13). One then has

$$S_\nu = \frac{A_\nu}{(1+z)^{3+s} r^2} , \qquad \begin{array}{l}\text{(radio source}\\ \text{with spectral} \quad (10.24)\\ \text{index } s)\end{array}$$

$$d_{L\nu} = (1+z)^{(3+s)/2} r.$$

An alternative way of describing apparent brightness uses the concept of *magnitude*. A source of apparent brightness S (or S_ν) is said to have an *apparent magnitude m* (or m_ν), where

$$m = -2.5 \log_{10} (S) + \text{constant},$$

$$(10.25)$$

$$m_\nu = -2.5 \log_{10} (S_\nu) + \text{constant}.$$

The number 2.5 arises from the convention that a change of 5 in apparent magnitude corresponds to a change by a factor of 100 in apparent brightness.

An *absolute magnitude M* (or M_ν) is also defined as the apparent magnitude which would be observed if the source were stationary at a standard distance D, usually taken to be 10 parsecs. Considering only bolometric quantities and combining eqns (10.18) and (10.25),

$$M = -2.5 \log_{10} (A/D^2) + \text{constant},$$

and subtracting this from eqn (10.25) gives

$$m - M = 2.5 \log_{10} (A/SD^2)$$

$$= 2.5 \log_{10} (d_L^2/D^2)$$

$$= 5 \log_{10} (d_L/D) \qquad\qquad (10.26)$$

where eqn (10.16) has been used. The relation between apparent magnitude and luminosity distance is therefore simple; however, the luminosity distance is more suited to the purposes of this book.

10.5. THE DENSITY TRANSFORMATION

This effect, which appears to have been discussed only recently [2] arises from the time taken for light to travel. If a photograph taken by O shows two galaxies G, G_1 to be at distances r, $r + \Delta r$ respectively, then, allowing for the time

light takes to travel, O will correctly infer that he is seeing G as it was at time $t = t_0 - r/c$, and G_1 as it was at time
$t_1 = t_0 - \frac{(r+\Delta r)}{c}$. The situation is illustrated in Fig. 10.5
for the case where O, G, G_1 are in a straight line.

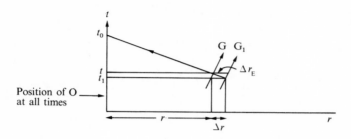

FIG. 10.5   Space—time diagram for the calculation of the density transformation in O's reference frame.

In this case Δr is the distance between G and G_1 as *observed* by O. But Δr is not a true distance, because it results
from comparing the position of G at time t with the position
of G_1 at the different time t_1. The true distance results
from comparing positions at the *same* time t, and this distance
is marked Δr_E.

Evidently the difference between Δr_E and Δr is just the
distance moved by G_1 in the time $(t-t_1)$. Hence, if G_1's speed
of recession is v_1, then

$$\Delta r_E = \Delta r + v_1(t-t_1) = \Delta r + v_1\Delta r/c = \Delta r(1+v_1/c).$$

Therefore if G_1 is at nearly the same distance as G and therefore, by the Hubble law, has nearly the same speed of recession v,

$$\frac{\Delta r_1}{\Delta r} = 1 + \frac{v}{c} = 1 + z. \tag{10.27}$$

Now suppose that a certain set of galaxies is photographed by O at time t_0, and that the photographed positions
of these galaxies approximately fill a rectangular volume of

dimensions Δr, Δy, Δz (G and G_1 are at two corners of the
volume). The true positions of the galaxies at time t will
fill a volume of dimensions Δr_E, Δy, Δz (distances at right-
angles to the velocity of galaxies are unchanged by the effect
under discussion). If n is the number of galaxies in the set,
the *density* of galaxies observed by O is

$$N = \frac{n}{\Delta r \; \Delta y \; \Delta z}$$

and the true density at time t is

$$N_E = \frac{n}{\Delta r_E \; \Delta y \; \Delta z}$$

so that, using eqns (10.5) and (10.27),

$$\frac{N}{N_E} = \frac{\Delta r_E}{\Delta r} = 1 + z. \tag{10.28}$$

The foreshortening of distance for the observer, described by
eqn (10.27), leads to an apparent density in excess of the
true density measured at the emitter. Note that, although
Δr and N are observed quantities, the suffice O has been sup-
pressed for simplicity, and to avoid confusion with the present density N_0, first introduced in Section 2.4.

10.6. CONSEQUENCES OF A SPECIAL RELATIVISTIC THEORY OF LIGHT PROPAGATION

It was mentioned at the start of this chapter that special
relativity (which we frequently abbreviate as SR) provides a
model of light propagation which avoids the problems of an
ether. It does so by postulating that the speed of light is
c relative to all observers. This entails a complete revision
of classical ideas of space and time, which has the effect
that many quantities hitherto thought to be absolute in nature,
such as the length of an object and the time interval between
two events, are seen to depend on the state of motion of the
observer who measures them. The consequences of this theory
for the optical effects already described are discussed in
Appendix A, and can be summarized as follows:

(i) The relation between red-shift z and speed of recession v is

$$1 + z = \left\{\frac{1 + v/c}{1 - v/c}\right\}^{\frac{1}{2}} \text{ (SR)}. \tag{10.29}$$

(ii) The factors f_{rs}, f_a, f_v, and N_r/N_{rE} are the same functions of z as in the classical wave theory (see Table 10.1), and the eqns (10.21), (10.24) for luminosity distance are also the same.

An important difference between these theories of light propagation is apparent from the behaviour of these expressions as the speed v of a receding galaxy approaches the speed of light c (Fig. 10.6). In the classical wave theory z, f_{rs}, f_a approach the finite values 1, 0.25, 0.25 and no difficulty arises as one considers greater values of v. According to special relativity, on the other hand, z becomes infinite and f_{rs} and f_a approach zero, implying that a galaxy with $v = c$ would have zero apparent brightness and so be unobservable. Speeds greater than c are also ruled out as they lead to imaginary values of z. Thus special relativity implies that only sources with $v < c$ are observable.

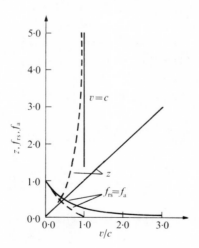

FIG. 10.6. Red-shift (z) and aberration factors $f_{rs} = f_a$ as a function of speed v for classical (solid line) and special relativistic (dashed) theories of light propagation.

In almost all cosmological models the restriction $v < c$ implies an upper limit on the distance at which sources can be observed. The observer O is effectively at the centre of a spherical surface of radius r_h, known as a *horizon*, and can observe only sources inside the horizon. For the simple cosmological models of Sections (2.4) to (2.6) the horizon distance r_h is found simply by substituting $v = c$ in eqns (2.8), (2.15) and (2.24) with the results:

$$r_h = \begin{cases} ct_H & \text{(steady state)} \\ 0.5\ ct_H & \text{(Milne)} \\ 0.4\ ct_H & \text{(Einstein–de Sitter)} \end{cases} \qquad \text{(SR)} \qquad (10.30)$$

where t_H is the Hubble time. In these models, and in most others, the horizon distance is of the same order as the distance ct_H, that is to say of the order 10^{10} light years at the present time.

The existence of a horizon or 'edge' to the observable universe may suggest the possibility of observing and enumerating *all* the galaxies within the horizon. There are a number of reasons why the possibility should not be taken too seriously. Firstly, galaxies near the horizon will be much fainter, because of the factors f_{rs} and f_a, than their distance alone would suggest. Secondly, the number of galaxies to be observed within the horizon is the integral

$$N_{total} = \int_0^{r_h} 4\pi r^2 N\ dr \qquad (10.31)$$

$$= \int_0^{r_h} 4\pi r^2 (1+z) N_E\ dr$$

where the factor $N/N_E = 1 + z$ has been introduced from eqn (10.28). Since $z \to \infty$ as $r \to r_h$, and since in most cosmologies N_E increases with r, the integral may well diverge at $r = r_h$, leading to an infinite total number of galaxies within the horizon (see Problem 10.2).

A further complication arises if we consider also the
fact that time intervals in special relativity are different
for different observers. For instance, the number density
N_E was calculated for the Milne model (see eqn (2.17)) on the
basis that a galaxy at distance r is observed by light which
left it a time $(t_0 - r/c)$ after the 'big bang'. However, this
time interval $(t_0 - r/c)$ is calculated in O's reference frame,
whereas what is needed is clearly the time interval in the
galaxy's own reference frame. The result of making allowances
for this is, as shown in Appendix A, that eqn (2.17) is
replaced by

$$N_E = N_0 \left\{ \frac{ct_H}{ct_H - 2r} \right\}^{3/2} \quad \text{(Milne, SR)}. \quad (10.32)$$

The same complication arises for the Einstein—de Sitter model,
but the required correction is then much more complicated
because the relative velocities of galaxies vary with time,
and this correction will not be discussed here. The steady-
state model is unaffected since N_E is not a function of time
in that model.

10.7. CONSEQUENCES OF A GENERAL RELATIVISTIC THEORY OF LIGHT PROPAGATION

The classical and special relativistic models of light propagation already discussed have one property in common; the
motion of a photon once emitted is not affected by the motion
of the galaxies or by the changing scale factor of the universe. In general relativity (hereafter often abbreviated to
GR) the situation is different; the gravitational field of any
massive body or of the universe as a whole is described as a
curvature of space-time which affects the motion of photons as
well as of material particles.

The difference can best be described in terms of the
dimensionless distance coordinates s_i introduced in eqn
(4.12b). We can describe the position of any particle (including a photon) relative to the Earth at time t either by conventional spherical polar coordinates (r, ϑ, φ) or by
dimensionless coordinates (s, ϑ, φ). Here ϑ, φ are angles

which define the *direction* in space of the particle's position, r is the particle's distance and s is defined by

$$r = R(t)s. \tag{10.33}$$

Notice that (s, ϑ, φ) are *comoving* coordinates; they are constant in time for any particle which moves with the general expansion of the universe.

Consider now the motion of a photon coming directly towards Earth. For such a photon ϑ and φ are constant, and r and s are decreasing. There are two obvious possibilities: (i) the photon's speed relative to us is c, and therefore

$$\mathrm{d}r/\mathrm{d}t = -c \text{ (equation of motion of a photon,}$$
$$\text{classical and SR optics)}; \tag{10.34}$$

(ii) the photon's speed is c relative to a local observer moving with the expansion of the universe. In this case the comoving coordinates must be used, giving

$$R(t)\ \mathrm{d}s/\mathrm{d}t = -c \text{ (equation of motion of a photon,}$$
$$\text{GR optics, flat space)}. \tag{10.35}$$

Differentiating eqn (10.33) and using eqn (10.35), one finds

$$\frac{\mathrm{d}r}{\mathrm{d}t} = -c + s\frac{\mathrm{d}R(t)}{\mathrm{d}t} \equiv -c + Hr \text{ (equation of motion}$$
$$\text{of a photon, GR optics,}$$
$$\text{flat space)} \tag{10.36}$$

which makes explicit the difference between the results of (i) and (ii).

General relativity involves a further correction resulting from the curvature of space. In a uniform continuum model this correction results in replacing eqn (10.35) by [3]

$$\frac{R(t)\mathrm{d}s/\mathrm{d}t}{\sqrt{(1-ks^2)}} = -c \text{ (equation of motion of a photon,}$$
$$\text{GR optics)} \tag{10.37}$$

where k is the *curvature index* which takes the value 0 for
flat space and ± 1 for curved space. It replaces in general
relativity the quantity k introduced in eqn (4.14), which
determines the sign of the total energy of a Newtonian system.

The coordinate s has been scaled in such a way that instead of a parameter describing the *amount* of curvature of
space one has a number k describing only the *sign* of the curvature. In fact the introduction to cosmology given in this
book is made possible by the complete identity of two cosmological differential equations: that of general relativity
with its k, and that of Newtonian cosmology with its k.

The reader may by now be uneasy on at least two counts.
First, eqn (10.36) seems to ascribe to a photon a speed different from c, and second, the whole of this section assumes
the existence of a 'universal time scale' in that the single
time coordinate t is used. On both counts it seems that
special relativity is violated. In fact this is not so;
special relativity says that the speed of light *as measured by
any observer* is c, and the quantity $(\mathrm{d}r/\mathrm{d}t)$ of eqn (10.36) is
not directly measurable. As regards the time scale, it is
true that differently moving observers will in general measure
time differently, but it can be shown [4] that a set of *comoving* observers in a Hubble law universe can agree on a common time scale. In all discussions of general relativistic
optics from now on, the symbol t will denote this common time.

It is now possible to derive the red-shift and associated
quantities in GR. Re-write eqn (10.37) as

$$\frac{\mathrm{d}s}{\sqrt{(1-ks^2)}} = -\frac{c\ \mathrm{d}t}{R(t)} \qquad \text{(equation of motion of a photon, GR)} \qquad (10.38)$$

and assume that this photon was emitted at time t_E from a
galaxy at coordinate s_E, and that the photon has coordinate
s_p at some later time t_p. Then integrating eqn (10.38) gives

$$\int_{s_E}^{s_p} \frac{\mathrm{d}s}{\sqrt{(1-ks^2)}} = -\int_{t_E}^{t_p} \frac{c\ \mathrm{d}t}{R(t)}$$

or

$$c \int_{t_E}^{t_p} \frac{\mathrm{d}t}{R(t)} = f(s_E) - f(s_p) \qquad (10.39)$$

where

$$f(s) \equiv \int_0^s \frac{\mathrm{d}s'}{\sqrt{(1-k(s')^2)}} = \begin{cases} \sin^{-1} s & \text{if } k = +1 \\ s & \text{if } k = 0 \\ \sinh^{-1} s & \text{if } k = -1. \end{cases} \qquad (10.40)$$

In particular, if the photon is observed (i.e., reaches $s = 0$) at time t_0 then

$$c \int_{t_E}^{t_0} \frac{\mathrm{d}t}{R(t)} = f(s_E). \qquad (10.41)$$

Now s_E, being the comoving coordinate of a galaxy, is constant in time. Hence if a second photon (or wavecrest) is emitted at time $t_E + \Delta t_E$ and observed at time $t_0 + \Delta t_0$, eqn (10.41) is replaced by

$$f(s_E) = c \int_{t_E+\Delta t_E}^{t_0+\Delta t_0} \frac{\mathrm{d}t}{R(t)} \sim c \int_{t_E}^{t_0} \frac{\mathrm{d}t}{R\,t} + \frac{c\Delta t_0}{R(t_0)} - \frac{c\Delta t_E}{R(t_E)} \qquad (10.42)$$

where it has been assumed that $R(t)$ does not vary significantly during the time intervals Δt_0, Δt_E.

Comparing eqn (10.42) with eqn (10.41) gives the very simple result

$$\frac{\Delta t_0}{R(t_0)} = \frac{\Delta t_E}{R(t_E)} ,$$

and as in Section (10.2) we infer that the red-shift z satisfies

$$1 + z = \frac{\Delta t_0}{\Delta t_E} = \frac{R(t_0)}{R(t_E)} . \qquad (10.43)$$

This result has a particularly pleasing interpretation in terms of wavelengths. If λ_E and λ_0 are emitted and observed wavelengths of a photon, then

$$\frac{\lambda_0}{\lambda_E} = \frac{c/\nu_0}{c/\nu_E} = \frac{\nu_E}{\nu_0} = 1 + z = \frac{R(t_0)}{R(t_E)} \qquad (10.44)$$

so that the wavelength has been 'stretched' in the course of the photon's journey by the same relative amount as the scale factor of the universe has increased.

It has already been noted (p. 162) that eqns (10.7) and (10.13) for the factors f_{rs} and f_ν remain valid in GR optics. This is also true for the aberration factor f_a, as is best shown by calculating the apparent brightness of a source directly, as follows.

Consider a source with total energy output L and intrinsic brightness A in the notation of Section 10.4, and suppose a system of comoving coordinates (s, ϑ, φ) has the *source* at its centre. By the cosmological principle the laws of optics must be the same in this coordinate system as in that based on O, so that photons must travel outwards from the source in straight lines. The distance these photons travel before reaching O at time t_0 is (by eqn (10.33)) $R(t_0)s$, so that the photons which strike unit area at O form a fraction $1/4\pi\{R(t_0)s\}^2$ of all those emitted. Combining this with the red-shift factor $f_{rs} = (1+z)^{-2}$ of eqn (10.7) gives the bolometric apparent brightness as

$$S = \frac{f_{rs}L}{4\pi\{R(t_0)s\}^2} = \frac{A}{(1+z)^2 R^2(t_0)s^2}$$

and the bolometric luminosity distance therefore as

$$d_L \equiv \left(\frac{A}{S}\right)^{\frac{1}{2}} = (1+z)R(t_0)s. \qquad (10.45)$$

This result looks different from that of eqn (10.21). However, the distance at which O observes this source is its distance at the time t_E determined by eqn (10.41). If we call this distance r, then by eqn (10.33) $r = R(t_E)s$, so that eqn (10.45) can be written

$$d_L = (1+z)\frac{R(t_0)}{R(t_E)}r . \qquad (10.46)$$

Finally, the expression (10.43) for the red-shift gives

$$d_L = (1+z)^2 r \qquad (10.47)$$

so that the form of the classical expression for luminosity distance is retained in general relativity. It follows by eqn (10.21) that the aberration factor f_a must also retain its classical form (eqn (10.15)).

10.8. THE CASE OF SMALL RED-SHIFTS FOR THE THREE THEORIES OF LIGHT PROPAGATION

In this section we show that despite the very different appearances of the three expressions for z in Table 10.2, they all reduce for close and slowly receding sources to

$$z \sim v/c \quad (v/c \ll 1 \text{ or } |t_E - t_0| \text{ sufficiently small}). \quad (10.48)$$

In the classical wave theory, eqn (10.48) is exact. In special relativity one can see from the second row of Table 10.2 that

$$\frac{v}{cz} = \frac{2 + z}{2 + 2z + z^2} \sim 1 \text{ for } z \ll 1 \text{ and therefore for } v/c \ll 1.$$

In the general relativistic case notice first that eqn (10.43) implies

$$R(t_0) = R(t_E)(1+z). \qquad (10.49)$$

Now expand $R(t_0)$ as a Taylor series in $(t_0 - t_E)$, retaining only the first two terms:

$$R(t_0) \approx R(t_E) + (t_0 - t_E)\left(\frac{dR}{dt}\right)_{t_E}$$

$$= R(t_E)\{1 + H(t_E)(t_0 - t_E)\}. \qquad (10.50)$$

Notice also that if $|t_0 - t_E|$ is small, eqn (10.41) implies that s_E is small and that

$$s_E \approx f(s_E) \approx c(t_0 - t_E)/R(t_E). \qquad (10.51)$$

Comparing eqn (10.49) with eqn (10.50) and using eqn (10.51) gives finally

$$z \approx H(t_E)(t_0 - t_E) \approx H(t_E)R(t_E)s_E/c$$

$$= H(t_E)r_E/c = v/c,$$

where $r_E \equiv R(t_E)s_E$ is the distance at which the source is observed.

10.9. THE DEFINITION AND MEASUREMENT OF DISTANCE

As general relativity involves the idea of curved space it is natural to ask whether the common-sense idea of distance is adequate in a cosmological context. The purpose of this section is to discuss possible ways of measuring and defining distance. The types of measurement to be considered all involve light signals (or signals using other parts of the electromagnetic spectrum) and can be classified as follows:

(i) Indirect methods, including measurement of apparent brightness and measurement of apparent angular diameter.

(ii) Direct methods, including parallax measurements and radar measurements.

(iii) Simultaneous measurements by widely spaced observers.

The indirect methods are those which require knowledge of some property of the object whose distance is to be measured. For instance, if one measures the apparent brightness S and the red-shift z of a distant galaxy, and if one knows its intrinsic brightness A, its distance r can be inferred from eqn (10.20). This is valid for all theories of light propagation if in the general relativistic case r is the distance defined by eqn (10.33) at the time of emission t_E:

$$r \equiv R(t_E)s. \qquad (10.52)$$

The apparent angular diameter of a source is defined as the greatest angle δ subtended at the observer by any pair of

light rays from the source (see Fig. 10.7). The figure is
valid in curved space since the light rays involved travel
radially, and therefore in straight lines, in the observer's
coordinate system. If the true diameter D of the source is
known, one can infer r from the geometry of the figure, which
gives $D = 2r \tan (\delta/2)$, or in the limit of small δ, $D = r\delta$.
This is also valid in general relativity [5] if r is defined
by eqn (10.52).

FIG. 10.7 Apparent diameter of a source.

The direct methods of measuring distance can be used with
no knowledge of the object's properties. The parallax method
involves measuring the change in direction of the line of
sight to the distant object as the Earth moves from one side
of its orbit to the other. In practice, this direction also
changes because of aberration of light resulting from the
Earth's changing velocity, but the amount of aberration is
known and can be corrected for. Fig. 10.8(a) illustrates the
light paths for classical or special relativistic optics (i.e.
in Euclidean space) and Fig. 10.8(b) does the same for general
relativity. The angle ϑ is called the *parallax* of the source.
It is evident from Fig. 10.8(a) that if r_E denotes the diameter of the Earth's orbit, then in Euclidean space $r_E = r \tan \vartheta$,
or since ϑ is always very small, $r = r_E/\vartheta$. It is conventional
to define a *parallax distance* d_p accordingly, so that in all
cases

$$d_p \equiv r_E/\vartheta; \quad d_p = r \text{ in Euclidean space.} \qquad (10.53)$$

The general relativistic calculation of d_p is given in Weinberg [6], and in our notation results in

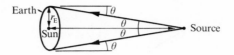

FIG. 10.8(a)   Parallax of a source in classical and special relativistic optics.

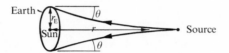

FIG. 10.8(b)   Parallax of a source in general relativistic optics.

$$d_p = \frac{R(t_0)s}{\sqrt{(1-ks^2)}} = \frac{(1+z)r}{\sqrt{(1-ks^2)}} \quad (G.R.) \tag{10.54}$$

where eqns (10.43) and (10.52) have been used.

The radar measurement of distance uses the following procedure: a short pulse of electromagnetic waves is emitted by the observer O at time t_0. The pulse is reflected from a distant object G and some of it returns to O at time t_1. It is then inferred that the pulse was reflected at a time

$$t_r \equiv \frac{t_0 + t_1}{2} \tag{10.55}$$

and that the distance of G from O at that time was the *radar distance*

$$d_r \equiv \frac{c}{2}(t_1 - t_0). \tag{10.56}$$

The reasoning behind this procedure is apparent from Fig. 10.9(a) which shows the motion of the pulse relative to O according to classical or special relativistic optics, for which the equation of motion of the pulse is $dr/dt = \pm c$.

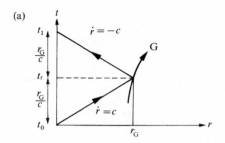

FIG. 10.9(a) Motion of a radar pulse in classical and special relativistic optics.

Notice that the motion of G (here shown moving away from O) is irrelevant, and that the radar distance d_r is in this case the same as the actual distance r_G at the time t_r.

In general relativity with flat space the equation of motion is, by an obvious extension of eqn (10.36), $dr/dt = \pm c + Hr$. As a result Fig. 10.9(a) must be replaced by Fig. 10.9(b). This figure is drawn for the steady-state model, in which H is constant, but the curvature of the radar pulse paths will be similar in all expanding-universe models. Note that t_{rp} denotes the actual time of reflection of the radar pulse, which is not the same as t_r. The equations of motion can easily be integrated for the case of constant H (see Problem 10.1) with the result that the radar distance is

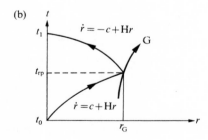

FIG. 10.9(b) Motion of a radar pulse in general relativistic optics for a steady-state universe.

$$d_r \text{ (steady state)} = \frac{ct_H}{2} \log_e \left\{\frac{1+r_G/ct_H}{1-r_G/ct_H}\right\} \tag{10.57}$$

Writing r for r_G for brevity, and expanding eqn (10.57) as a Taylor series in r/ct_H, one finds

$$d_r \text{ (steady state)} = ct_H\left\{\frac{r}{ct_H}+\frac{1}{3}\left(\frac{r}{ct_H}\right)^3+\frac{1}{5}\left(\frac{r}{ct_H}\right)^5+ \ldots\right\}$$

$$= r\left\{1+\frac{1}{3}\left(\frac{r}{ct_H}\right)^2+\frac{1}{5}\left(\frac{r}{ct_H}\right)^4+ \ldots\right\} \tag{10.58}$$

so that radar distances approximate 'true' distances closely at short range ($r \ll ct_H$).

It is apparent from eqns (10.54) and (10.58) that both the 'direct' methods of distance measurement produce results which differ from the distance r defined by eqn (10.48), but also that these three quantities are nearly equal at short distances. This suggests that one might approach the definition of distance by dividing a distance such as OG into a large number n of equal sections, each section being so short that the different measures of distance would be approximately equal. The limit $n \to \infty$ would then give a unique measure of distance. For instance, if n observers O, O_2, ..., O_n, equally spaced along the line OG, each make radar measurements so synchronized that the pulse reflection time t_{rp} is the same for each, then each individual radar distance would in the steady-state theory be like eqn (10.58) with r replaced by r/n. Adding these n distances together would then give

$$d_r \text{ (steady state, } n \text{ observers)} = n\left(\frac{r}{n}\right)\left\{1+\frac{1}{3}\left(\frac{r}{nct_H}\right)^2+ \ldots\right\}$$

$$= r\left\{1+\frac{1}{3}\left(\frac{r}{nct_H}\right)^2+ \ldots\right\}. \tag{10.59}$$

Taking the limit of $n \to \infty$ gives a measure known as the *proper distance*,

$$d_{prop} \text{ (steady state)} = r. \tag{10.60}$$

This simple result is a consequence of the fact that the steady-state model has flat space, $k = 0$. The corresponding result for general k is [7]

$$d_{\text{prop}} \text{ (at time } t_E) = R(t_E) \int_0^s \frac{ds'}{\sqrt{(1-k(s')^2)}} \equiv R(t_E)f(s) \tag{10.61}$$

where $f(s)$ is the function first introduced in eqn (10.40). It follows that the proper distance of a source at the time of emission t_E equals the distance r of eqn (10.48) if, and only if, $k = 0$.

Obviously the set of cooperating observers envisaged in eqn (10.59) does not exist on a cosmological scale, so that d_{prop} is not a directly measurable quantity. The radar and parallax methods are also limited in range — the radar method to within the solar system and the parallax method at present to a few tens of light years — so that the only practical measures of cosmological distances at present are those based on inferences from apparent brightness and apparent diameter. Both are subject to a basic uncertainty in that the inference depends in each case on knowing the intrinsic brightness or diameter. Also the intrinsic brightness may change with time, as discussed in more detail in Section 11.5, while the measurement of apparent diameter is complicated by the fact that galaxies do not have sharp edges. A comparison of the two measures can help to reduce the uncertainties; however *in principle* the two need not be considered separately since they are simply related in all models:

$$d_L:d \text{ (apparent diameter)} = (1+z)^2 r:r = (1+z)^2:1.$$

The next chapter therefore concentrates on the notion of apparent brightness. Apparent diameters are discussed in Section 11.6.

Of the distances r, d_p, d_r, d_L, and d_{prop} introduced in this chapter, by eqns (10.33) and (10.61) both r and d_{prop} satisfy the Hubble law in the form

$$\frac{dr}{dt} = Hr \text{ with } H \equiv \mathring{R}/R.$$

PROBLEMS

(10.1)   Show that if a radio source has a spectrum of the type
(10.12) with $s = 1$, then $E(\nu_1,\nu_2)$ as defined in (10.8)
depends only on the ratio ν_2/ν_1. Verify that this
implies $f_\nu = 1$, in agreement with (10.13) for $s = 1$.

(10.2)   Using eqns (10.29) to (10.32), (B.23), and (2.8) to
(2.9), show that for a special relativistic model of
light propagation the number of galaxies within the
horizon is
(a)   infinite for the Milne model
(b)   finite, and less than $8\sqrt{2}\ \pi\ N_0\ (ct_H)^3$, for the
steady-state model.
(Note: the change of variable $u \equiv 0.5 - r/ct_H$ is helpful
in (a), and similarly $w \equiv 1 - r/ct_H$ is helpful in (b).)

(10.3)   By integrating the equations of motion illustrated in
Fig. 10.9(b), and using the definition (10.56), verify
the expression (10.57) for a radar distance in the
steady-state theory.

11

OPTICAL OBSERVATIONS AND COSMOLOGICAL MODELS

11.1. OBSERVED QUANTITIES

The models of the universe which have been discussed so far can be summarized as follows. Sources of electromagnetic radiation are distributed more or less uniformly through space and are moving according to Hubble's law. The scale factor $R(t)$ obeys a differential equation with three parameters C, λ, k. For a large number of the sources one can measure the apparent brightness S (or S_ν) and the red-shift z. Within this framework there are several possibilities for relating observed quantities to cosmological parameters. One can observe, to quote only the more important items:

(i)   the apparent brightness of sources as a function of their red-shifts;

(iia) the number of sources having apparent brightnesses within a given range, or

(iib) the number of sources having red-shifts within a given range;

(iii) the background brightness of the night sky.

In each of these cases one can, with some additional assumptions, predict the observed quantities theoretically for a particular set of values of (C, λ, k). The process of prediction is illustrated in this chapter for the simpler cosmological models. The brightness-red-shift relation (i) is discussed in Section 11.2, the source counts (iia) in Section 11.3, and the background brightness (iii) in Section 11.4.

This account suggests that the basic questions of cosmology are near solution — one has only to compare accurate observational data with the predictions and the values of C, λ, k can be found. That this has not been done is a result partly of observational limitations — for instance the correction factor f_ν of Section 10.2 is particularly difficult to estimate for photographic observations of galaxies at large red-shift, and it is just these observations which are needed

to distinguish between cosmological models, as noted in Section 11.2. A more fundamental problem is the uncertainty in
the intrinsic brightness of sources. This uncertainty has
two components.

First, different *types* of galaxy have different intrinsic brightness at the present time, and presumably at other
times, so that a plot of observed apparent brightness against
red-shift will show considerable scatter. The scatter can be
reduced by looking at clusters of galaxies and recording only
the brightest galaxy of each cluster, since such brightest
members are found empirically to have more nearly equal intrinsic brightnesses than the general population. In recent
times there has been a tendency to use the fifth brightest
member of a cluster instead, and also to use criteria based
on the type of galaxy.

Secondly, the intrinsic brightness of each galaxy must
change during its lifetime, since this brightness must be
small both before stars have formed and after most stars have
exhausted their nuclear fuel. In all big-bang models the
galaxies are formed in a single relatively short period of
time. Thus, if two otherwise identical galaxies are observed at different distances, the differing light travel
times mean that we see them at different stages of their history and therefore probably with different intrinsic brightnesses. The amount of this evolutionary change in brightness and of the correction needed for it are very uncertain
at present (see Section 11.5).

In the next three sections of this chapter attention is
concentrated on the three cosmological models first mentioned
in sections 2.4, 2.5, and 2.6 — the steady-state (S), Milne
(M), and Einstein—de Sitter (E) models. These models have
the advantage of mathematical simplicity, and also yield
simple values of the deceleration parameter $q = -1$ (steady
state), 0 (Milne), and +0.5 (Einstein—de Sitter). One can
consider each cosmological model in conjunction with any of
the three theories of light propagation, leading to a 3×3
matrix of possibilities with an associated simple notation, as
in Table 11.1. This notation is used frequently in the

following sections.

TABLE 11.1

Notation for various models

		Theories of light propagation		
		Classical Wave Theory (C)	Special Relativity (SR)	General Relativity (GR)
Cosmological models	Steady State (S)	SC	SSR	SGR
	Milne (M)	MC	MSR	MGR
	Einstein— de Sitter (E)	EC	ESR	EGR

11.2. THE APPARENT BRIGHTNESS — RED-SHIFT RELATION

In this section the luminosity distance d_L introduced in
Section 10.4 is used as the most convenient measure of apparent brightness. The procedure for calculating d_L as a function of red-shift z is different for different theories of
light propagation.

For the classical and special relativistic theories the
most convenient expression for d_L is contained in Table 10.1
and is repeated here for convenience:

$$d_L = (1+z)^2 r. \tag{11.1}$$

These theories also provide unique relations between the red-
shift of a galaxy and its speed of recession v, and these are
reproduced here from Table 10.2:

$$v = cz \quad (C),$$

$$v = c \frac{(1+z)^2 - 1}{(1+z)^2 + 1} = \frac{cz(2+z)}{(1+z)^2 + 1} \quad (SR). \tag{11.2}$$

Finally, the travel time of a photon from a galaxy observed at
distance r is simply r/c in both theories. The relations
between r and v, calculated on this basis in eqns (2.8),

(2.15), and (2.24) for the simpler cosmological models, are therefore valid and are reproduced here

$$r = v \, t_H \qquad (S)$$

$$r = \frac{v \, t_H}{1 + v/c} \qquad (M) \qquad\qquad (11.3)$$

$$r = \frac{2v \, t_H}{2 + 3v/c} \qquad (E)$$

where in each case t_H is the Hubble time, the reciprocal of the present value H_0 of the Hubble parameter.

The relation between d_L and z is now easily found by combining the appropriate part of eqn (11.3) with the appropriate part of eqn (11.2) and substituting the result in eqn (11.1). For instance, if one is interested in the Milne model and a special relativistic theory of light propagation, one combines the result (M) of eqn (11.3) with the result (SR) of eqn (11.2) and finds

$$r = ct_H \frac{v/c}{1+v/c} = \frac{ct_H\{(1+z)^2-1\}/\{(1+z)^2+1\}}{1+\{(1+z)^2-1\}/\{(1+z)^2+1\}}$$

$$= \frac{ct_H\{(1+z)^2-1\}}{\{(1+z)^2+1\} + \{(1+z)^2-1\}} = \frac{ct_H(2+z)z}{2(1+z)^2} \quad (MSR).$$

Substituting this in eqn (11.1) gives

$$d_L = ct_H \, z(1+0.5z) \quad (MSR). \qquad (11.4)$$

This, and the five corresponding results arising from different choices of models, are summarized in the first two rows of Table 11.2.

For a general relativistic theory of light propagation the procedure is necessarily quite different. Since the travel time of a photon is not simply r/c, the results (11.3) are not valid, and it is convenient to replace eqn (11.1) by eqn (10.45):

TABLE 11.2.

Luminosity distance in units of ct_H for various models

	Steady-state model	Milne model	Einstein–de Sitter model
Classical wave optics	$z(1+z)^2$ (SC)	$z(1+z)$ (MC)	$\dfrac{z(1+z)^2}{1+1.5z}$ (EC)
Special relativity	$\dfrac{z(2+z)(1+z)^2}{(1+z)^2+1}$ (SSR)	$z(1+0.5z)^*$ (MSR)	$\dfrac{z(1.0.5z)(1+z)^2}{1+2.5z+1.25z^2}$ (ESR)
General relativity	$z(1+z)$ (SGR)	$z(1+0.5z)^*$ (MGR)	$2\{(1+z)-(1+z)^{\frac{1}{2}}\}$ (EGR)

*Other observational implications are also identical (Table 11.3, equations (10.30) and (11.83)).

$$d_L = (1+z)\ R(t_0)s. \qquad (11.5)$$

The dimensionless coordinate s is itself related to the time of emission t_E by eqns (10.40) and (10.41) (the suffix on s has been dropped for convenience)

$$c\int_{t_E}^{t_0} \frac{\mathrm{d}t}{R(t)} = f(s) = \begin{cases} \sin^{-1} s & \text{if } k = +1 \\ s & \text{if } k = 0 \\ \sinh^{-1} s & \text{if } k = -1 \end{cases} \qquad (11.6)$$

and the time of emission is related to the red-shift by eqn (10.43):

$$1 + z = \frac{R(t_0)}{R(t_E)}\ . \qquad (11.7)$$

These equations allow one in principle to calculate d_L as a function of z, given the form of the function $R(t)$ and the value of k. The procedure is fairly complex and the three simple cosmological models will be considered separately.

For the steady-state model we know from eqn (2.10) that the scale factor $\bar{R}(t)$ is equal to $\exp\{H(t-t_0)\}$, and it can also be shown that $k = 0$ [1]. The re-scaling procedure of Section 4.5 allows an arbitrary choice of the re-scaling factor P when $k = 0$, and the most convenient choice is that which leads to

$$R(t) = \frac{c}{H} \exp(Ht) \quad (S).$$ (11.8)

It is also possible and convenient, since the steady-state model has no natural origin of time, to choose

$$t_0 = 0 \text{ and hence } R(t_0) = \frac{c}{H} \quad (S).$$ (11.9)

Substituting eqns (11.8) and (11.9) into (11.6) with $k = 0$, one finds

$$s = c \int_{t_E}^{0} \frac{\mathrm{d}t}{(c/H)\exp(Ht)} = H \int_{t_E}^{0} \exp(-Ht)\mathrm{d}t = \exp(-Ht_E) - 1$$

whence

$$\exp(-Ht_E) = 1 + s.$$ (11.10)

But from eqns (11.7), (11.8), and (11.9)

$$1 + z = \frac{R(t_0)}{R(t_E)} = \frac{c/H}{(c/H)\exp(Ht_E)} = \exp(-Ht_E).$$ (11.11)

Comparing eqns (11.10) and (11.11) gives $s = z$, and substituting this with eqn (11.9) into (11.5) gives

$$d_L = \frac{c}{H}(1+z)z = ct_H z(1+z) \quad (SGR).$$ (11.12)

The Einstein–de Sitter model is discussed in Section 6.2 and has $k = 0$ and $R(t) \propto t^{2/3}$. The scaling factor P is again arbitrary and will be chosen so that

$$R(t) = 3ct_0^{1/3} t^{2/3},$$ (11.13)

and in particular

$$R(t_0) = 3ct_0. \tag{11.14}$$

Substituting this in eqn (11.6) with $k = 0$ gives

$$s = c \int_{t_E}^{t_0} \frac{t^{-2/3} \, dt}{3ct_0^{1/3}} = \frac{1}{t_0^{1/3}} (t_0^{1/3} - t_E^{1/3})$$

$$= 1 - (t_E/t_0)^{1/3}$$

whence

$$t_E = t_0 (1-s)^3. \tag{11.15}$$

Also, using eqn (11.15),

$$1 + z = \frac{R(t_0)}{R(t_E)} = \frac{t_0^{2/3}}{t_E^{2/3}} = (1-s)^{-2}$$

whence

$$s = 1 - (1+z)^{-0.5}. \tag{11.16}$$

Substituting eqns (11.16) and (11.14) into eqn (11.5) gives finally

$$d_L = 3ct_0 (1+z)\{1 - (1+z)^{-0.5}\}$$

$$= 2ct_H\{1+z - (1+z)^{0.5}\} \quad \text{(EGR)}, \tag{11.17}$$

where the relation $t_H = 1.5 \, t_0$, first noted for the Einstein–de Sitter model in Section 2.6, has been used.

The Milne model is discussed in Section 6.2.6 and is defined by $k = -1$ and

$$R(t) = ct; \quad R(t_0) = ct_0. \tag{11.18}$$

Substituting into eqn (11.6) with $k = -1$ gives

$$\sinh^{-1} s = \int_{t_E}^{t_0} \frac{dt}{t} = \log_e (t_0/t_E);$$

$$t_E = t_0 \exp(-\sin h^{-1} s) = t_0 (\surd(s^2+1)+s)^{-1}.† \qquad (11.19)$$

The red-shift, using eqns (11.18) and (11.19), is given by

$$1 + z = \frac{R(t_0)}{R(t_E)} = \frac{t_0}{t_E} = \surd(s^2+1) + s$$

which can be solved† to give

$$s = \frac{(1+z)^2 - 1}{2(1+z)} . \qquad (11.20)$$

Finally, substituting eqns (11.20) and (11.18) into eqn (11.5) gives

$$d_L = \frac{ct_0}{2}\{(1+z)^2-1\} = ct_H z(1+0.5z) \quad (MGR), \qquad (11.21)$$

where eqn (2.13) has also been used.

The results of this section are summarized in Table 11.2 and illustrated in Figs 11.1 to 11.3. All three figures have two properties in common: first, all curves lie below the straight line $z = d_L/ct_H$, and secondly the value of z for given d_L increases as one goes from the steady-state model to big-bang models of increasing deceleration parameters. The first property is a simple consequence of $d(d_L/ct_H)/dz > 1$, which holds for all entries in Table 11.2. The second property follows from the fact that for a given observed distance, the velocity of recession is higher for models with a large deceleration parameter. It is evident from the figures that all models predict $d_L \sim ct_H z$ for $z \ll 1$, and this allows obser-

---

†Suppose $y \equiv \surd(s^2+1)+s$. Then $\surd(s^2+1) = y - s$, $s^2 + 1 = y^2 + s^2 - 2sy$, $y^2 - 2sy = 1$, $s = (y^2-1)/2y = \frac{1}{2}(y-y^{-1})$. With $y = 1 + z$ this verifies eqn (11.20). Also, if $y \equiv e^u$ then $s = \frac{1}{2}(e^u-e^{-u}) \equiv \sinh u$, so $\exp(\sinh^{-1} s) = \exp(u) = y = \surd(s^2+1) + s$, as required for eqn (11.19).

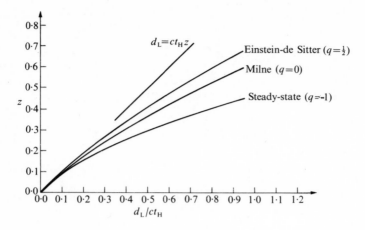

FIG. 11.1 Luminosity distance as a function of red-shift according to classical optics using the equations of Table 11.2.

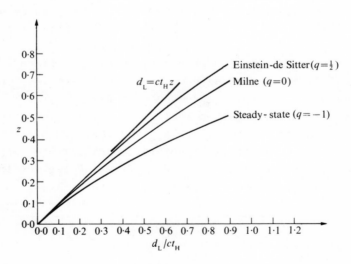

FIG. 11.2 Luminosity distance as a function of red-shift according to special relativistic optics using the equations of Table 11.2.

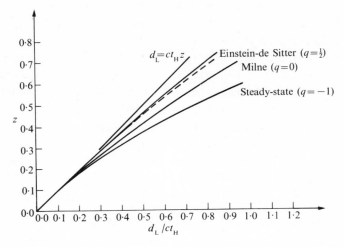

FIG. 11.3 Luminosity distance as a function of red-shift according to
general relativistic optics using the equations of Table 11.2. The dashed
curve is calculated from the power series approximation, eqn (11.72).

vational estimates of H_0 to be made which are independent of
the model chosen. It is possible to distinguish clearly between models only at rather large red-shifts, where the observational uncertainties are also large. Data of apparent
brightness versus red-shift therefore do not yet discriminate
very convincingly between different cosmological models, as
will also appear in Section 11.5.

11.3. SOURCE COUNTS

It is possible, as noted in Section 11.1, to consider either
the number of sources within specified brightness limits or
the number within limiting values of red-shift. However, the
source count technique has been used primarily for radio
sources, and red-shifts cannot be measured for these unless
they can be identified with a visible or photographable object. It is therefore usual to consider the quantity $N(S_\nu)$,
defined as the total number of sources with apparent radio
brightness greater than some limiting value S_ν, as a function
of S_ν. In this section the bolometric brightness S is considered for simplicity, and the calculation of $N(S)$ is discussed with the simplying assumption that all sources at whatever distance have the same intrinsic brightness A. A more

accurate calculation [2] would involve an integration over the range of possible values of A at any given time, together with an allowance for evolution.

Since the intrinsic brightness of radio sources not identified with visible objects is poorly known, theoretical values of S are uncertain, and attention has been concentrated on the *shape* of the $N(S)-S$ graph. In this context it is interesting to consider what shape the graph would have in a uniform static universe with flat space (i.e. Euclidean geo-metry). In such a universe the space density of sources is equal at all times to its present value N_0, so that the number in a sphere of radius r_L centred on the observer is

$$N(r_L) = \frac{4\pi}{3}N_0 r_L^3.$$

Since there is no red-shift in a static universe the apparent bolometric brightness of a source on the surface of the sphere is, as in eqn (10.19), $S = A/r_L^2$. Sources *in* the sphere all have apparent brightnesses greater than S, so that $N(r_L)$ is the same as $N(S)$. It follows that

$$N(S) = \frac{4\pi}{3}N_0 r_L^3 = \frac{4\pi}{3}N_0\left(\frac{A}{S}\right)^{3/2} \equiv KS^{-3/2}$$

and therefore

$$\log\{N(S)\} = K' - (3/2)\log(S) \quad \text{(uniform, static, Euclidean universe)} \quad (11.22)$$

where K and K' involve only the constant intrinsic brightness and constant space density of the sources. Thus a graph of $\log\{N(S)\}$ against $\log(S)$ would in this very special case be a straight line with a slope of $(-3/2)$.

When discussing source counts in an expanding universe it is necessary to specify a theory of light propagation. For the classical and special relativistic theories it is reason-able to assume Euclidean geometry and to consider again a sphere of radius r_L centred on the observer. The number of sources observed in the sphere (i.e. observed at distances up to r_L) is

$$N(r_L) \equiv N(S) = 4\pi \int_0^{r_L} r^2 N \mathrm{d}r \qquad (11.33)$$

where N, the observed source density at distance r, is related
to the true density N_E by eqn (10.28), so that

$$N(S) = 4\pi \int_0^{r_L} r^2 (1+z) N_E \mathrm{d}r. \qquad (11.34)$$

The limiting brightness S corresponding to the distance r_L is
given by eqn (10.20) so that

$$S = \frac{A}{(1+z_L)^4 r_L^2} , \qquad (11.35)$$

z_L being the red-shift corresponding to distance r_L.

It is now necessary to specify the theory of light propagation and the cosmological model. These determine z_L as a
function of r_L, and z and N_E as functions of r, so that $N(S)$
and S can both be found as functions of r_L. It would, of
course, be desirable to solve eqn (11.35) for r_L as a function
of S, but this is not possible for most models. The best that
can be done is to calculate $N(S)$ and S separately as functions
of r_L, and plot one against the other. The results needed are
the expressions for z as a function of v in Table 10.1, and the
pressions for z as a function of v in Table 10.1, and the
results of Sections 2.4 to 2.6 reproduced below:

$$v/c = x \,(\mathrm{S}), \qquad x/(1-x) \,(\mathrm{M}), \qquad \frac{x}{(1-1.5x)}(\mathrm{E}) \qquad (11.36)$$

and

$$N_E = \begin{cases} N_0 \;(\mathrm{S}), & N_0/(1-x)^3 \;(\mathrm{MC}), \\[2ex] N_0/(1-2x)^{3/2} \;(\mathrm{MSR}), & N_0/(1-1.5x)^2 \;(\mathrm{EC}), \end{cases} \qquad (11.37)$$

where the notation

$$x \equiv r/ct_H \qquad (11.38)$$

has been introduced. It is convenient to rewrite eqns (11.34) and (11.35) in terms of x and $x_L \equiv r_L/ct_H$ as

$$N(S) = 4\pi(ct_H)^3 \int_0^{x_L} x^2(1+z)N_E dx,   \tag{11.39}$$

$$S = \frac{A/(ct_H)^2}{(1+z_L)^4 x_L^2}.   \tag{11.40}$$

The calculation is now straightforward except possibly for the evaluation of the integral. For instance, if one is interested in the steady-state model and a classical theory of light propagation, Table 10.1 with eqn (11.36) (S) gives

$$z = v/c = x, \quad z_L = x_L   \text{(SC)}$$

and so eqns (11.39) and (11.40) become respectively, using eqn (11.37) (S),

$$N(S) = 4\pi(ct_H)^3 \int_0^{x_L} x^2(1+x)N_0 dx = 4\pi(ct_H)^3 N_0 \left(\frac{x_L^3}{3}+\frac{x_L^4}{4}\right)   \text{(SC)}   \tag{11.41}$$

and

$$S = \frac{A/(ct_H)^2}{(1+x_L)^4 x_L^2}   \text{(SC)}.   \tag{11.42}$$

The resulting graph is shown in Fig. 11.4. Corresponding graphs for models SSR, MC, MSR, and EC are also shown in Figs 11.4 and 11.5 and some of the underlying mathematics is discussed in Problem 11.2.

All curves shown have a negative slope simply because the number of sources down to brightness S has been plotted, and this number must decrease as S is increased. There are two main effects which displace the curves shown relative to the line $N \propto S^{-3/2}$ appropriate to the static universe. The first is that expansion weakens the apparent brightness of any given source, and this displaces the curves to the left. Furthermore this effect increases with distance and this results in the curvature shown. The second effect explains the difference between the steady-state curves and those for the big-bang models. It is this: as one observes lower apparent brightnesses, one is observing earlier states of the universe, which correspond in big-bang models to higher densities. The curves for the big-bang models are therefore expected to lie above the corresponding curve for the steady-state model.

The general relativistic calculation of source counts differs from the foregoing because the photon travel time is no longer r/c, and also because in non-Euclidean space the volume between distances r and $r + dr$ differs from the Euclidean value $4\pi r^2 dr$. In terms of the dimensionless distance coordinate s, the number of sources observed at coordinates between s and $s + ds$ is [3]

$$dN = 4\pi R^3(t_E)N(t_E)\frac{s^2 ds}{\sqrt{(1-ks^2)}} \qquad (11.43)$$

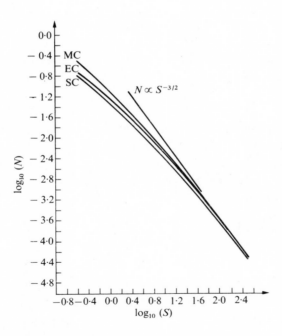

FIG. 11.4 Number $N(S)$ of sources with apparent brightness greater than S, according to classical optics using eqns (11.41) and (11.42) and the results of Problem 11.2.

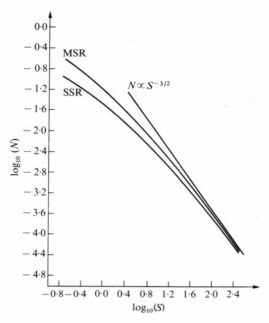

FIG. 11.5 Number $N(S)$ of sources with apparent brightness greater than S, according to special relativistic optics using the results of Problem 11.2.

where t_E is the time of emission from a source which is observed at coordinate s, and N_E has been re-named $N(t_E)$ to make the time dependence explicit. The time t_E and the coordinate S are, of course, related by the photon travel time eqns (10.41) and (10.40):

$$c \int_{t_E}^{t_0} \frac{\mathrm{d}t}{R(t)} = f(s) = \int_0^s \frac{\mathrm{d}s'}{\sqrt{(1-k(s')^2)}} . \qquad (11.44)$$

In the context of a source count, the time of observation t_0 can be regarded as constant. Thus a change from s to $s + \mathrm{d}s$ involves a change from t_E to $t_E + \mathrm{d}t_E$, where

$$c \int_{t_E + \mathrm{d}t_E}^{t_0} \frac{\mathrm{d}t}{R(t)} = \int_0^{s+\mathrm{d}s} \frac{\mathrm{d}s'}{\sqrt{(1-k(s')^2)}} \qquad (11.45)$$

Subtracting eqn (11.45) from eqn (11.44) gives

$$- \frac{c \, dt_E}{R(t_E)} = \frac{ds}{\sqrt{(1-ks^2)}} \, ,$$

and using this to change the variable in eqn (11.43) from s to t_E gives the simpler result

$$dN = -4\pi \, cR^2(t_E)N(t_E)s^2dt_E \qquad (11.46)$$

where s must now be regarded as a function of t_E determined by eqn (11.44).

The source count $N(S)$ is now obtained by integrating eqn (11.46) from the present time t_0 back to a limiting time t_L which is the time corresponding to an apparent brightness S. The steady-state model has $N(t_E) = N_0$, and so

$$N(S) = 4\pi c N_0 \int_{t_L}^{t_0} R^2(t_E)s^2 dt_E \quad \text{(SGR)}. \qquad (11.47)$$

In the Milne and Einstein—de Sitter models, assuming no creation or destruction of sources during the interval of cosmic time (t_L, t_0), one has instead $N(t_E) = N_0\{R(t_0)/R(t_E)\}^3$ from eqn (2.16), and so

$$N(S) = 4\pi c N_0 R^3(t_0) \int_{t_L}^{t_0} \frac{s^2 dt_E}{R(t_E)} \quad \text{(MGR, EGR)}. \qquad (11.48)$$

The corresponding limiting value of S is found from eqn (10.45) as

$$S = \frac{A}{R^2(t_0)(1+z_L)^2 s_L^2} \qquad (11.49)$$

where s_L, z_L are values of s, z corresponding to the time t_L.

The calculation for each model can now proceed using the relations between s, z, R, and t_E derived in Section 11.2. For the Milne model we have from eqns (11.18), (11.20), and the relation $t_0 = t_H$ that

$$R(t_E) = ct_E; \quad s = \frac{(t_H/t_E)^2 - 1}{2(t_H/t_E)}; \quad 1 + z = \frac{t_H}{t_E}. \quad (11.50)$$

Substituting the first two of these in eqn (11.48) with $t_0 = t_H$ gives

$$N(S) = 4\pi c N_0 (ct_H)^3 \int_{t_L}^{t_H} \left\{ \frac{(t_H/t_E)^2 - 1}{2(t_H/t_E)} \right\}^2 \frac{dt_E}{ct_E} \quad (11.51)$$

A change of variable to $y \equiv t_E/t_H$, $y_L \equiv t_L/t_H$ converts this to

$$N(S) = \pi N_0 (ct_H)^3 \int_{y_L}^{1} \left\{ \frac{1}{y} - y \right\}^2 \frac{dy}{y}$$

$$= \pi N_0 (ct_H)^3 \int_{y_L}^{1} \{ y^{-3} - 2y^{-1} + y \} \, dy$$

$$= \pi N_0 (ct_H)^3 \{ 0.5(y_L^{-2} - 1) + 2 \log_e(y_L) + 0.5(1 - y_L^2) \}$$

and if we define $w \equiv 1/y^2 = (t_H/t_E)^2$, $w_L \equiv 1/y_L^2 = (t_H/t_L)^2$, this reduces to

$$N(S) = \pi N_0 (ct_H)^3 \left\{ \frac{w_L}{2} - \log_e(w_L) - \frac{1}{2w_L} \right\}. \quad (11.52)$$

The limiting value of S is found from eqn (11.49). Notice that from eqn (11.50), $R(t_0) = ct_0 = ct_H$ and also

$$(1+z)s = \{ (t_H/t_E)^2 - 1 \}/2 = (w-1)/2,$$

so that

$$S = \frac{4A}{(ct_H)^2 (w_L - 1)^2} \quad \text{(MGR)}. \quad (11.53)$$

Equations (11.52, 53) are equivalent to $N(S)$ in problem (11.2) with the replacement $w_L \rightarrow 1/(1-2x_L)$. The calculations of $N(S)$ and S for the models SGR and EGR are equally straightforward; the results are given in Problems 11.3 and 11.4 and the graphs for all three models are shown in Fig. 11.6. The qualitative features of these curves have already been discussed in connection with Figs 11.4 and 11.5

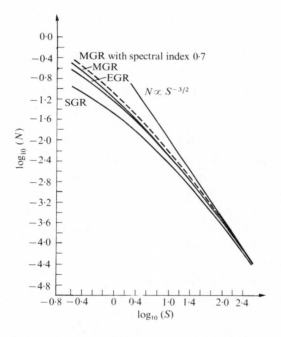

FIG. 11.6 Number $N(S)$ of sources with apparent brightness greater than S, according to general relativistic optics using eqns (11.52) and (11.53) and the results of Problems 11.3 and 11.4. The dashed curve relates to a radio apparent brightness S_ν, assuming a spectral index 0.7, not to a bolometric apparent brightness S.

The dotted curve represents a source count for the SGR model in which the apparent *radio* brightness S_ν is considered rather than the bolometric brightness S, assuming a spectral index of 0.7. It can be derived from the corresponding curve for S as follows. By eqns (10.22) and (10.24) the ratio S_ν/S

for a source with spectral index 0.7 and red-shift z_L is

$$S_\nu/S = (A_\nu/A)(1+z_L)^{0.3} .$$

Now $N(S)$, calculated from eqn (11.48) or (11.49), is simply
the number of sources observed at red-shifts up to z_L, and can
equally well be denoted by $N(S_\nu)$. The procedure for convert-
ing a graph of $N(S)$ against S into a graph of $N(S_\nu)$ against S_ν
is therefore simple: multiply each value of S by
$(A_\nu/A)(1+z_L)^{0.3}$. Since the graph shown is logarithmic and we
are interested only in its shape, the constant factor A_ν/A can
be ignored, and this has been done in the figure.

It is now possible to compare the theoretical source
counts with observation. The theoretical curves all have
slopes which are less than the value (-1.5) for the straight
line $N(S) \propto S^{-3/2}$. However, the observed source counts [4]
lead to curves which are generally *steeper* than this line,
with slopes of about (-1.8). The discrepancy is so large that
it almost forces the conclusion that evolutionary effects are
important, in the sense that either the intrinsic brightness
of radio sources changes with time or their space density
changes with time in a way different from that allowed for
by eqns (11.48) or (11.49). Since the steady-state theory
cannot allow either form of evolution, the source counts are
generally considered to be strong evidence against that theory.
It is also clear that the source counts cannot be used to
decide between Friedmann models until evolutionary effects are
well understood.

11.4. HOW DARK IS THE NIGHT SKY? OLBERS' PARADOX
(a) Qualitative discussion
The preceding two sections discussed observations which can
be made only using a fairly large optical or radio telescope.
No instruments at all are needed, however, to make the observation that the sky is dark at night! The surprising fact
that this rules out some otherwise possible models of the
universe was pointed out by de Cheseaux [5] in 1744 and independently by Olbers [6] in 1826. The result is now commonly

known as the Olbers' paradox [7].

The paradox arises when one considers a universe which
is

   (i)   infinite in extent;
  (ii)   uniform in space — the mean density of galaxies in
          space does not vary with distance;
 (iii)   static — there are no systematic motions of the galax-
          ies, also the mean brightness of galaxies does not
          alter with time;
  (iv)   infinitely old.

Such a model universe differs from any commonly accepted to-
day in having, by property (iii), no overall expansion.
Property (iv) in conjunction with (iii) is also ruled out by
the conservation of energy, since it implies that each galaxy
has emitted an infinite amount of energy in the past. In
Olbers' time, however, the first three properties at least
seemed plausible and were quite widely accepted, with the word
'star' replacing 'galaxy'. The fact that property (iv) is
essential to the paradox was not noted explicitly by Olbers.

A non-mathematical statement of the paradox can be given
as follows. Since stars are of finite size, any straight
line drawn through an infinite space containing a uniform
density of stars must sooner or later intersect the surface of
a star. More precisely, if p is the probability that a line
of length L intersects the surface of at least one star (p is
certainly not zero if L is much greater than the average distance between stars) then the probability that a line of
length nL does *not* intersect the surface of any star is $(1-p)^n$,
and tends to zero as $n \to \infty$. Thus a line of sight, drawn in
any direction from an observer, must end on the surface of a
star. Since the apparent brightness of a luminous surface at
rest does not vary with distance, the sky should appear in all
directions as bright as the surface of a typical star.

The paradox can be resolved in several ways. Evidently a
finite universe (lacking property (i)), or one in which the
density of stars or galaxies decreases rapidly with distance
from the observer (lacking property (ii)), can avoid the problem. So can a universe infinite in extent but finite in age

(i.e. lacking property (iv)), since if the stars were all
created a time t_0 ago no star at a distance greater than ct_0
can now be seen. However, the resolution of the problem which
is currently of most interest is that which discards property
(iii) and replaces the static universe by an expanding one.
The expansion introduces three effects: (a) the light from
each galaxy is weakened by the red-shift and aberration factors (10.7) and (10.15), (b) the apparent density of galaxies
in space is increased by the factor (10.28), and (c) the true
density of galaxies in space at large distances is increased
since they are observed at times earlier and nearer the start
of the expansion than the present. Since (a) decreases the
light received from distant sources while (b) and (c) increase
it, a more detailed calculation is needed to find whether, in
a particular expanding model universe, the paradox is in fact
resolved.

(b) Quantitative discussion
The detailed calculation can make use of the theory of source
counts in the following way. If one writes the number of
sources within a sphere of radius r_L formally as

$$N(r_L) = \int_0^{r_L} \frac{\mathrm{d}N}{\mathrm{d}r} \, \mathrm{d}r$$

then the total energy received by a unit area at 0 from these
sources is

$$E(r_L) = \int_0^{r_L} S\frac{\mathrm{d}N}{\mathrm{d}r} \, \mathrm{d}r \qquad (11.54)$$

where S is the average bolometric apparent brightness of the
$(\mathrm{d}N/\mathrm{d}r)\mathrm{d}r$ sources between r and $r + \mathrm{d}r$. (The unit area must
be arranged so as to accept energy from all directions, i.e.
it must be a sphere of unit cross-section.) The total energy
received from all sources is obtained simply by making $r_L \to \infty$
(if no horizon exists) or $r_L \to r_h$ (if a horizon exists at r_h).

One thus has the following recipe: to convert an expression of
the type (11.34) for a source count $N(S)$ into an expression
for the total energy E incident on unit area, multiply the
integrand by S and extend the upper limit of integration to
the horizon. The reader should satisfy himself that the same
procedure is reasonable for the general relativistic expressions (11.47) and (11.48), except that the $lower$ limit of
integration is extended to $t_E = 0$ (for MGR, EGR) or to $t_E = -\infty$
(for SGR). The relevant expression for S is of course A/d_L^2,
with d_L given by eqn (10.21) or eqn (10.45) depending on the
theory of light propagation in use, and with A taken to be an
average value of intrinsic brightness.

In a universe satisfying the conditions (i) to (iv), and
also having flat space, one has $S = A/r^2$ and $dN/dr = 4\pi N_0 r^2$.
There is also no horizon, so that the total energy E is eqn
(11.54) with $r_L = \infty$, or

$$ E = \int_0^\infty \frac{A}{r^2} \cdot 4\pi N_0 r^2 dr = \infty \qquad (11.55) $$

Eqns (11.54) and (11.55) overestimate E, by neglecting
the fact that some fraction f_r of the light from stars at distance r will be intercepted by nearer stars and prevented from
reaching the observer. Allowance for this effect can be made
by multiplying the integrand of each equation by a correction
factor $(1-f_r)$. One would expect that f_r will be small for
small r and approach unity for large r, so that the effect of
the correction on eqn (11.55) will be to impose a 'cut-off' at
some large but finite r, and so make E for the infinite static
universe finite rather than infinite. The correction is less
important for the expanding universe models discussed below,
and will in fact be neglected. This is justifiable because
most of the light in such models comes from distances less than
or of the order of ct_H, both because of the existence of hori-
zons in some models and because in any case the red-shift and
aberration factors of Table 10.1 reduce the apparent brightness of more distant sources.

We can estimate f_r as follows. A cylinder of cross-sectional area A and length r, stretching from the observer to a star S, will on average contain a number $n = NAr$ of stars, N being the average number density of stars in the universe. The probability that any one of these stars will eclipse S is (a/A), a being the typical cross-section of a star. The probability that *none* of the n stars will eclipse S is therefore $1 - f_r = (1-a/A)^n \sim 1-na/A$, so that $f_r \sim na/A = Nar$. The estimates $N \sim 10^{-12}$ stars per cubic light year and $a \sim 10^{-13}$ square light years then lead to $f_r \sim 10^{-25}r$. Since ct_H is about 2×10^{10} light years, it follows that $f_r \lesssim 10^{-15}$ for $r \lesssim ct_H$. f_r can therefore be neglected in expanding universe models. A similar calculation involving galaxies rather than stars would greatly overestimate f_r, because the outer layers of galaxies are transparent. The corresponding numbers for such a calculation are $N \sim 4 \times 10^{-21}$ and $a \sim 10^9$, leading to $f_r < 10^{-1}$ for $r \lesssim ct_H$.

Note that a universe which satisfied (ii) to (iv) in flat space, but was a finite sphere of radius r_u centred on the observer, would give

$$E = 4\pi A N_0 \int_0^{r_u} dr = 4\pi A N_0 r_u .$$

We can therefore define, for any model universe, an equivalent radius $r(\text{equiv.})$, which is the radius of a finite uniform static universe which would have the same value of E, by

$$r(\text{equiv.}) \equiv E/4\pi A N_0. \qquad (11.56)$$

To illustrate the calculation of E, consider first the steady-state model with classical optics. The source count integral is, from eqn (11.41),

$$4\pi(ct_H)^3 \int_0^{x_L} x^2(1+x)N_0 \, dx$$

with $x \equiv v/c = r/ct_H$ in the steady-state model. Our recipe requires us to multiply the integrand by

$$S = \frac{A}{(ct_H)^2 x^2 (1+x)^4}$$

(see eqn (11.42)) and integrate out to the horizon. In classical optics the horizon is at infinity, so the result is

$$E = 4\pi A N_0 c t_H \int_0^\infty \frac{x^2(1+x)}{x^2(1+x)^4} \, dx \quad (SC)$$

$$= 4\pi A n_0 c t_H \int_0^\infty \frac{dx}{(1+x)^3} = 2\pi A N_0 c t_H. \quad (11.57)$$

In terms of the equivalent radius of eqn (11.56), this gives

$$r(\text{equiv.}) = 0.5 \, ct_H. \quad (SC) \quad (11.58)$$

As a second illustration, consider the Milne model with general relativistic optics. The source-count integral (11.51) must have its integrand multiplied by S and the lower limit of integration must be zero. From eqns (10.45) and (11.50), and using also that $R(t_0) = ct_H$ for the Milne model, we find

$$S = \frac{A}{(ct_H)^2 (1+z)^2 s^2} = \frac{4A}{(ct_H)^2 \{(t_H/t_E)^2 - 1\}^2}.$$

Multiplying the integrand of eqn (11.51) by this and setting $t_L = 0$ gives

$$E = 4\pi N_0 c t_H \int_0^{t_H} \frac{4A}{[2(t_H/t_E)]^2} \frac{dt_E}{t_E} = 4\pi A N_0 c t_H \int_0^{t_H} \frac{t_E dt_E}{(t_H)^2}$$

$$= 2\pi A N_0 c t_H \quad (MGR) \quad (11.59)$$

with a corresponding equivalent radius

$$r(\text{equiv.}) = 0.5 \; ct_{\text{H}}. \qquad \text{(MGR)} \qquad (11.60)$$

Similar results can be calculated for other models, and the results are shown in Table 11.3, details of the calculation being discussed in Problems 11.5—11.7. Model ESR has been excluded from the table, as it was from the discussion of source counts, because of the difficulty of evaluating the correction discussed at the end of Section 10.6.

TABLE 11.3

Equivalent radii of model universes

| | | | Theories of light propagation | |
		Classical	Special relativity	General relativity
Cosmological model	Steady-state	0.5	0.2876	0.25
	Milne	1.0	0.5	0.5
	Einstein—de Sitter	0.5		0.4

The table shows equivalent radii $r(\text{equiv.})$, related to the total energy incident from the night sky by eqn (11.56), in units of ct_{H}.

11.5. THE LUMINOSITY DISTANCE AS A POWER SERIES IN THE RED-SHIFT

The preceding three sections have given exact results for three particularly simple cosmological models. When one considers more complicated models it is usually necessary either to analyse the problem numerically, for particular values of the parameters, or to seek an approximate method of solution. One such approximate method involves the expansion of the scale factor $R(t_{\text{E}})$ as a power series in the photon travel time or 'time of flight' defined as

$$t_{\text{f}} \equiv t_0 - t_{\text{E}}. \qquad (11.61)$$

Such an expansion is most accurate for nearby objects which have short photon travel times. It is therefore particularly appropriate for calculating the luminosity distance d_L, which itself is most accurately known for neaby objects. The present section uses the power series approach to examine the relation between d_L and the red-shift z. In this context the notation $O(x^n)$, meaning 'terms containing x^n and higher powers of x' is useful. Also the symbol $\approx$ will be used to indicate that powers of x or z higher than those explicitly given are neglected.

The starting points are the expressions (10.43) for z and (10.44) for d_L. Expanding $R(t_E)$ as a Taylor series about the present time t_0 gives

$$R(t_E) = R(t_0) + \left[\frac{dR}{dt}\right]_{t_0} (t_E - t_0) + \tfrac{1}{2}\left[\frac{d^2R}{dt^2}\right]_{t_0} (t_E - t_0)^2 + \dots \quad (11.62)$$

From eqns (4.19) and (4.22) it follows that

$$\left[\frac{dR}{dt}\right]_{t_0} = H_0 R(t_0) \text{ and } \left[\frac{d^2R}{dt^2}\right]_{t_0} = - H_0{}^2 q_0 R(t_0). \quad (11.63)$$

Substituting these results and eqn (11.61) into eqn (11.62) gives

$$R(t_E) = R(t_0)\{1 - H_0 t_f - \tfrac{1}{2} q_0 H_0{}^2 t_f{}^2 + O(t_f{}^3)\}, \quad (11.64)$$

and if one introduces a new variable

$$x \equiv H_0 t_f, \quad (11.65)$$

this is equivalent to

$$R(t_E) \approx R(t_0)\{1 - x - \tfrac{1}{2} q_0 x^2\}, \quad (11.66)$$

and so by eqn (10.43) to

$$1 + z = \frac{R(t_0)}{R(t_E)} \approx \{1 - x - \tfrac{1}{2} q_0 x^2\}^{-1}. \quad (11.67)$$

It is now necessary to decide what level of accuracy is needed. We already know from Section 11.2 that $d_L \sim c t_H z$ for small z. A significant improvement on this will involve the second power of z and be of the form $d_L \sim c t_H z (1 + \alpha z + \dots)$. We therefore need to retain terms in z^2 or equivalently in x^2, and shall neglect higher powers. Using the general result $(1-y)^{-1} = 1 + y + y^2 + \ldots$ in eqn (11.67) gives

$$z = (x+\tfrac{1}{2}q_0 x^2) + (x+\tfrac{1}{2}q_0 x^2)^2 + \ldots$$

$$\approx x + (1+\tfrac{1}{2}q_0)x^2. \tag{11.68}$$

This result can be inverted[†] to give

$$x \approx z - (1+\tfrac{1}{2}q_0)z^2. \tag{11.69}$$

Since eqn (10.44) involves the coordinate s, we now need a relation between s and x or z. The starting point here is eqn (11.44). The right-hand side of this equation can be approximated as

$$f(s) = \int_0^s (1-k(s')^2)^{-\frac{1}{2}} \, ds'$$

$$= \int_0^s \{1+\frac{k}{2}(s')^2+\ldots\} \, ds'$$

$$= s + \frac{ks^3}{6} + \ldots = s + O(s^3), \tag{11.70}$$

so that $f(s) \approx s$ if powers higher than s^2 are neglected. Also, if we define

$$x' \equiv H_0(t_0-t) = (t_0-t)/t_{\mathrm{H}} \tag{11.71}$$

then replacing t_{E} by t and hence x by x' in eqn (11.68) gives

$$\frac{1}{R(t)} = \frac{1}{R(t_0)}\{1+x'+(1+\tfrac{1}{2}q_0)(x')^2+\ldots\}$$

[†]If $z = x + \alpha x^2 + \ldots$, and if $x = z - \beta z^2 + \ldots$ then $z = z-\beta z^2+\alpha(z-\beta z^2)^2 = z + (\alpha-\beta)z^2 + \ldots$ which is correct up to terms in z^2 if $\beta = \alpha$.

so that the left-hand side of eqn (11.44) is

$$c \int_{t_E}^{t_0} \frac{dt}{R(t)} \equiv ct_H \int_0^x \frac{dx'}{R(t)} = \frac{ct_H}{R(t_0)} \int_0^x \{1+x'+(1+\tfrac{1}{2}q_0)(x')^2+\ldots\} \; dx'$$

$$\approx \frac{ct_H}{R(t_0)}\left(x+\frac{x^2}{2}\right).$$

Combining this with eqn (11.70) gives

$$s \approx \frac{ct_H}{R(t_0)}\left(x+\frac{x^2}{2}\right)$$

and substituting this in eqn (11.44) gives

$$d_L = ct_H(1+z)\left\{x+\frac{x^2}{2}+\ldots\right\}$$

$$= ct_H(1+z)\left\{z-(1+\tfrac{1}{2}q_0)z^2+\frac{z^2}{2}+O(z^3)\right\} \quad \text{(using eqn (11.69))}$$

$$\approx ct_H z\{1+0.5(1-q_0)z\}. \tag{11.72}$$

This is the required power series for d_L as a function of z. By comparing it with observations it is possible to estimate both the Hubble parameter $H_0 \equiv 1/t_H$ and the deceleration parameter q_0, and this is one of the better ways of estimating these quantities.

The accuracy of the estimate of q_0 can be improved by making allowances for the evolution in brightness of sources. Define an evolutionary parameter α_e which is related to the rate of change of the intrinsic brightness A of a source by

$$\alpha_e \equiv \frac{1}{H_0 A_0}\left.\left(\frac{dA}{dt}\right)\right|_{t_0} \tag{11.73}$$

where A_0 is the present intrinsic brightness. Then the intrinsic brightness at the time of observation t_E is

$$A(t_E) = A(t_0) + \left[\frac{dA}{dt}\right]_{t_0} (t_E - t_0) \cdots$$

$$\equiv A_0 - \left[\frac{dA}{dt}\right]_{t_0} t_f + \cdots$$

$$= A_0\{1 - \alpha_e H_0 t_f + \cdots\} \equiv A_0\{1 - \alpha_e x + O(x^2)\} \qquad (11.74)$$

where eqns (11.61) and (11.65) have been used. Now the apparent brightness S of the source must satisfy

$$\frac{A(t_E)}{d_L^2} = S = \frac{A_0}{(d_L^*)^2}$$

where d_L^* is the luminosity distance which would be inferred from the apparent brightness *without* allowing for evolution. Hence

$$d_L^* = d_L\{A_0/A(t_E)\}^{\frac{1}{2}} = d_L\{1 - \alpha_e x + O(x^2)\}^{-\frac{1}{2}}$$

$$= d_L\{1 + 0.5\alpha_e x + O(x^2)\}$$

$$= d_L\{1 + 0.5\alpha_e z + O(z^2)\} \qquad (11.75)$$

which with eqn (11.72) implies

$$d_L^* \approx ct_H z\{1 + 0.5(1 + \alpha_e - q_0)z\}. \qquad (11.76)$$

One way of expressing this result is to say that evolution replaces q_0 by an *apparent* deceleration parameter

$$q_{0a} = q_0 - \alpha_e. \qquad (11.77)$$

Estimates of α_e vary considerably. Gott *et al.* [8] consider it to lie in the range from -0.4 to -1.2, which combined with Sandage's estimate [9] $q_{0a} = 1.0 \pm 1$ leads to $-1.2 < q_0 < +1.6$. The more recent estimate [10] $q_{0a} = -0.15 \pm 0.57$ suggests a negative value of q_0, implying by eqn (4.24) a positive value of the cosmological constant λ. This conclusion was also reached in a more detailed discussion by Gunn and Tinsley [11] but must be regarded as tentative.

11.6. HORIZONS AND ANGULAR DIAMETERS IN GENERAL RELATIVISTIC MODELS

In discussing the existence of a horizon in a GR universe one has to specify which measure of distance is to be used. For instance, one could ask either

(i) Which sources observable at the present time have the greatest value of the dimensionless distance coordinate s?

or

(ii) Which of these sources had the greatest distance r at the time of emission of the light now being observed?

Since by eqn (10.37) $ds/dt < 0$ for all the relevant photons, the greatest possible value of s is found by going to the earliest possible time of emission — to $t_E = 0$ for models MGR, EGR and to $t_E = -\infty$ for model SGR. One then finds from Eqns (11.10), (11.15), and (11.19) that the greatest value of s is

$$s_{max} = \begin{cases} \infty & (SGR) \\ \infty & (MGR) \\ 1 & (EGR) \end{cases} \qquad (11.78)$$

and corresponds in each case to an infinite red-shift. Thus only the Einstein—de Sitter model has a definite horizon in the sense of question (i). The *present* proper distance of a source on this horizon is found from eqns (10.61) and (11.14) to be

$$d_{prop\ max}(t_0) = 3ct_0 = 2ct_H \text{ (EGR)}. \qquad (11.79)$$

Weinberg [12] gives results generalizing eqn (11.79) to different ($\lambda = 0$) model universes.

The greatest value of r, on the other hand, is not necessarily found at the earliest time, since eqn (10.36) shows that dr/dt need not be negative — the motion of a photon towards the observer may be more than balanced by the overall expansion of the universe. We must therefore consider each model in detail. For model SGR it follows from eqns (11.8), (11.10), and (11.11) that

$$r \equiv R(t_E)s = ct_H\{1-\exp(Ht_E)\} = ct_H z/(1+z) \quad \text{(SGR)}. \quad (11.80)$$

Similarly from eqns (11.13), (11.15), and (11.16)

$$r = 2ct_H\left\{\left(\frac{t_E}{t_0}\right)^{2/3} - \frac{t_E}{t_0}\right\} = 2ct_H\left\{\frac{1}{1+z} - \frac{1}{(1+z)^{3/2}}\right\} \quad \text{(EGR)} \quad (11.81)$$

and from eqns (11.18) to (11.20)

$$r = \frac{ct_H}{2}\left\{1-\left(\frac{t_E}{t_0}\right)^2\right\} = \frac{ct_H}{2}\left\{1-\frac{1}{(1+z)^2}\right\} \quad \text{(MGR)}. \quad (11.82)$$

Inspection of these results shows that for models SGR and MGR the greatest distance does correspond to an infinite red-shift, the values being

$$r_{max} = \begin{cases} ct_H & \text{(SGR)} \\ 0.5ct_H & \text{(MGR)}. \end{cases} \quad (11.83)$$

For model EGR, on the other hand, an infinite red-shift corresponds to $r = 0$. The greatest value of r is found at a red-shift such that $dr/dz = 0$, and is

$$r_{max} = \frac{8ct_H}{27} \quad \text{(EGR)}, \quad (11.84)$$

corresponding to a red-shift $z = 1.25$ and to a time of emission $t_E = t_0/(2.25)^{3/2} = 8t_0/27$.

The way in which model EGR differs from SGR and MGR can be seen more clearly from Fig. 11.7, which is based on eqns (11.80) to (11.82). Because gravity is the dominant force in model EGR, the recessional velocity $v = Hr$ of every galaxy is very high soon after the start of the expansion, and so photons emitted at early times at first traverse regions in which $Hr > c$ and are 'swept' away from the observer. They can reach the observer only when the rate of expansion has slowed down sufficiently. In models SGR and MGR, on the other hand, the

velocities of all galaxies are constant or decrease as one goes
further back in time, so no 'sweeping' of photons occurs.

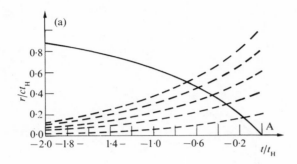

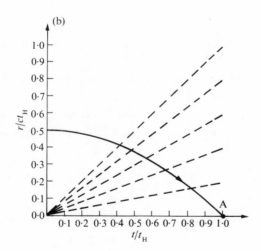

FIG. 11.7 Free photon propagation (solid lines) to reach the observer
at A, and motion of galaxies (dashed lines), for three general relativistic
cosmological models.   (a) Steady-state; the present time has been taken as
zero.   (b) Milne; a photon emitted at the big bang is shown.

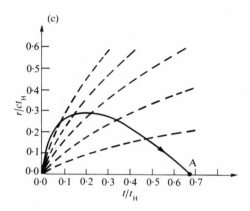

FIG. 11.7(c) Einstein—de Sitter; a photon emitted at the big bang is shown.

Because the angular diameter of a source at distance r is $\delta = D/r$, D being the true diameter, eqn (11.84) implies that a set of identical sources at different distances in an EGR universe will all have apparent angular diameters greater than or equal to a limiting value

$$\delta_{min} = \frac{D}{r_{max}} = \left(\frac{27}{8}\right)\frac{D}{ct_H} \quad (EGR).$$

Furthermore, the limiting value δ_{min} occurs not at infinite red-shift but at $z = 1.25$, less than many observed red-shifts. In fact all Friedmann models with $q > 0$ have such a limit for given D [13]. Thus observations of angular diameter as a function of red-shift or luminosity distance can provide evidence to distinguish between cosmological models.

Some recent measurements [14] suggest that $q_0 = 0.3 \pm 0.2$, thus apparently ruling out the steady-state model as discussed in this book.

PROBLEMS

(11.1) Using eqns (11.1) to (11.3), verify the first two rows
of Table 11.2.

(11.2) Dimensionless radio source counts and apparent brightness are defined by

$$N' = \frac{N(S)}{4\pi N_0 (ct_H)^3}; \quad S' = \frac{(ct_H)^2 S}{A} .$$

Using eqns (11.36) to (11.40) verify the following:
For model MC,

$$N' = \frac{1}{3}\left(\frac{x_L}{1-x_L}\right)^3 \text{ and } S' = \frac{(1-x_L)^4}{x_L^2} .$$

For model MSR,

$$N' = \frac{1}{8}\left\{\frac{1}{1-2x_L}+2\ln(1-2x_L)-(1-2x_L)\right\} \text{ and } S' = \left\{\frac{1-2x_L}{x_L}\right\}^2 .$$

For model SSR,

$$N' = \int_0^{x_L} x^2\left\{\frac{1+x}{1-x}\right\}^{\frac{1}{2}} dx \text{ and } S' = \left\{\frac{1-x_L}{x_L+x_L^2}\right\}^2 .$$

(11.3) With the notation of the previous question and eqns
(11.8) to (11.11) and (11.47) show that for model SGR,

$$N' = \frac{2}{w} - \frac{1}{2w^2} - 1.5 + \ln(w) \text{ and } S' = \frac{1}{w^2(w-1)^2}$$

with $w \equiv \exp(-Ht_L)$.

(11.4) Using eqns (11.13) to (11.15) and (11.48) show that
for model EGR,

$$N' = \frac{8}{3}(1-y_L)^3 \text{ and } S' = \left\{\frac{y_L}{2(1-y_L)}\right\}^2$$

with $y_L \equiv (t_L/t_0)^{1/3}$.

(11.5) Verify the entries in Table 11.3 for models MC and EC, using eqns (11.36) to (11.38) and noting that the greatest observable distances in these models are $ct_{\rm H}$ (MC) and $2ct_{\rm H}/3$ (EC), corresponding in each case to infinite velocities of recession.

(11.6) Verify the entries in Table 11.3 for models SSR and MSR, using eqns (11.36) to (11.38) and the horizon distances of eqn (10.30). Note that

$$\int_0^1 \left(\frac{1-x}{1+x}\right)^{3/2} \, \mathrm{d}x = 0.2876.$$

(11.7) Verify the entries in Table 11.3 for models SGR and EGR.

APPENDIX A

Effects of a special relativistic theory of light propagation

A1. THE LORENTZ TRANSFORMATION

In this Appendix the special relativistic results required in Section 10.6 are established.

Suppose that an observer O, at rest relative to the centre of our galaxy, and an observer O' at rest relative to the centre of a galaxy G, measure the position and time of an event E. Suppose also that they agree to choose Cartesian coordinate systems S and S' such that corresponding axes in the two systems are parallel, and that S' is receding from O along the x-axis of S with a constant speed v, as in Fig. A.1.

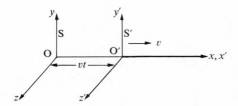

FIG. A.1   Relation between reference frames in relativistic motion, at time t in frame S.

Suppose finally that the observers agreed to synchronize their clocks to read zero at the instant when the two sets of axes coincided. Then Newtonian ideas of the nature of space and time lead to the following relations

$$
\begin{aligned}
x' &= x - vt \\
y' &= y \\
z' &= z \\
t' &= t
\end{aligned}
\qquad (A.1)
$$

where x, y, z are the space coordinates and t the time of any

event E, measured by O, and x', y', z', t' are corresponding
quantities measured by O'.

It is a consequence of (A.1) that the velocity of any
moving object or signal depends on the reference frame in
which it is measured. In particular, a signal propagating
with speed c relative to O along the x-axis will satisfy
$x = ct + x_0$, $y = z = 0$, and it follows from (A.1) that
$x' = (c-v)t' + x_0$, $y' = z' = 0$, so that the signal propagates
with speed $(c-v)$ relative to O'. There is, however, strong
experimental evidence that the speed of light in vacuum is the
same relative to all observers. If one accepts this constancy
of the speed of light as an axiom, and looks for the simplest
possible equivalent of (A.1) which will satisfy the axiom, one
finds the *Lorentz transformation*:

$$
\begin{aligned}
x' &= \beta(x-vt) \\
y' &= y \qquad \text{with } \beta \equiv (1-v^2/c^2)^{-\frac{1}{2}} \\
z' &= z \\
t' &= \beta\left(t-\frac{vx}{c^2}\right)
\end{aligned}
\tag{A.2}
$$

It is easy to verify that $x = ct$ implies $x' = ct'$ and conversely, so that a signal travelling along the x-axis with
speed c in one frame also has speed c in the other frame. It
follows from (A.17) below that the same is true for a signal
travelling in a general direction.

The replacement of (A.1) by the Lorentz transformation
(A.2) has far-reaching consequences for classical physics. In
this Appendix we confine ourselves to discussing effects of
direct relevance to observational cosmology as discussed in
this book.

A.2. TIME DILATION AND THE RED-SHIFT
Suppose that a light source at rest in S' emits light pulses
which are seen by both observers. Let the events E_1, E_2 be
the emission of two such pulses, and let the time interval
between them, measured by O', be $\Delta t'_E$. According to classical
physics the light pulses are seen by O at times separated by a
time interval $\Delta t_0 \neq \Delta t'_E$, because of the classical red-shift

of eqn (10.4), but O can allow for this red-shift and infer
that the time interval between E_1 and E_2 was $\Delta t_E = \Delta t_0/(1+v/c)$.
The inferred Δt_E would then be equal to $\Delta t'_E$ by (A.1). In
special relativity, however, this Δt_E is still a time interval
in O's reference frame S, and the Lorentz transformation must
be used to relate it to the time interval in S'. Neglecting
the y and z coordinates, the coordinates of E_1 and E_2 in S'
can be written (x', t') and $(x', t' + \Delta t'_E)$ respectively. If
the coordinates in S are (x, t) and $(x + \Delta x, t + \Delta t_E)$, then
(A.2) gives

$$x' = \beta(x-vt) = \beta(x+\Delta x-v(t+\Delta t_E));$$

$$t' = \beta\left(t-\frac{vx}{c^2}\right), \quad t'+\Delta t'_E = \beta\left(t+\Delta t_E-\frac{v}{c^2}(x+\Delta x)\right)$$

whence

$$\Delta x = v\Delta t_E$$

and

$$\Delta t'_E = \beta\left(\Delta t_E-\frac{v}{c^2}\Delta x\right) = \beta\left(1-\frac{v^2}{c^2}\right)\Delta t_E.$$

Recalling the definition of β in (A.2), this can finally be
written

$$\Delta t'_E = (1-v^2/c^2)^{\frac{1}{2}}\Delta t_E = \Delta t_E/\beta; \quad \Delta t_E = \beta\Delta t'_E. \qquad \text{(A.3)}$$

Although time intervals are thus seen to be relative to
the state of motion of an observer, the interval $\Delta t'_E$ has a
special importance in that it is measured by an observer for
whom the two events coincide in space. Thus a clock located
at the point in S' where E_1 and E_2 occur will actually register
a time interval $\Delta t'_E$ extending from E_1 to E_2. On the other
hand, there is no clock anywhere which directly registers the
time interval Δt_E; this time interval is inferred by O from
more complicated observations. For this reason $\Delta t'_E$ is called
a *proper* time interval. The fact that $\Delta t_E \neq \Delta t'_E$ is known as

the *time dilation* effect, and can be described in general as
follows:

> If E_1 and E_2 are events in the history of an object,
> the time interval between them, in a reference frame in
> which the object is moving with speed v, is greater by
> the factor β than the proper time interval between them.

One consequence of this is explored in Problem A.1. To
discuss the red-shift we return to equation (A.3), and note
that if E_1 and E_2 are emissions of successive wavecrests of
light, the *proper* frequency of emission is $\nu'_E \equiv 1/\Delta t'_E$, and
the *proper* frequency of arrival is $\nu_0 \equiv 1/\Delta t_0$. Hence the
definition (10.2) of the red-shift must be replaced by

$$1 + z = \frac{\nu'_E}{\nu_0} = \frac{\Delta t_0}{\Delta t'_E} . \tag{A.4}$$

Combining eqns (10.4) and (A.3) and (A.4), we find

$$1 + z = \frac{\Delta t_0}{\Delta t_E} \frac{\Delta t_E}{\Delta t'_E} = \beta\left(1+\frac{v}{c}\right) \equiv \frac{1+v/c}{(1-v^2/c^2)^{\frac{1}{2}}} , \tag{A.5}$$

and after some algebra

$$1 + z = \left\{\frac{1+v/c}{1-v/c}\right\}^{\frac{1}{2}} \tag{A.6}$$

thus confirming eqn (10.29) in Section 10.6. The arguments
leading to eqns (10.7) and (10.13), and those equations them-
selves, remain valid, with ν_E replaced by ν'_E and Δt_E by $\Delta t'_E$
throughout. For example,

$$f_{rs} = \frac{1}{(1+z)^2}$$

is an equation which holds in classical and special relativistic optics.

A.3. THE LORENTZ—FITZGERALD CONTRACTION AND THE DENSITY
TRANSFORMATION

Suppose that a thin rod, parallel to the x-axis, is at rest in
S' with one end at the origin of S'. Then the x-coordinates of
the ends of the rod in S' will be $x'_1 \equiv 0$ and $x'_2 \equiv L'$ at all

times, where L' is the length of the rod measured in S'. For
obvious reasons L' is called the proper length of the rod. By
the first equation of (A.2), these results imply $x_1 = vt$ and
$x_2 = vt + L'/\beta$ for the x-coordinates of the ends at a time t
in S. Thus if O measures the positions of both ends at the
same time in his reference frame S, the distance between these
positions will be L'/β, and he will regard this as a valid
measurement of the length L. Thus we have the result

$$L = L'/\beta \qquad\qquad (A.7)$$

which is known as the Lorentz—Fitzgerald contraction and can
be generalized as follows:

> The length (measured in the direction of motion) of
> any object moving with speed v relative to the observer
> is less by a factor β than the proper length of the ob-
> ject. Lengths perpendicular to the direction of motion
> are the same as corresponding proper lengths.

The application to the density transformation of Section
10.5 is clear. Regarding the galaxies G, G_1 as ends of a rod,
Δr_E of eqn (10.27) corresponds to L and the *proper* distance
$GG_1 = \Delta r'_E$ corresponds to the proper length L', so that
$\Delta r'_E = \beta \Delta r_E$. The quantity of interest is now the ratio of the
observed density N to the *proper* density $N'_E \equiv n/\Delta r'_E \Delta y' \Delta z'$.
Since the first half of eqn (10.27) remains valid, the argu-
ment leading to eqn (10.28) can be replaced by

$$\frac{N}{N'_E} = \frac{\Delta r'_E}{\Delta r} = \frac{\Delta r'_E}{\Delta r_E}\frac{\Delta r_E}{\Delta r} = \beta\left(1+\frac{v}{c}\right) = 1 + z \qquad (A.8)$$

where we have used (A.5) and also $\Delta y' = \Delta y$, $\Delta z' = \Delta z$ from
(A.2). Thus eqn (10.28) remains valid if one replaces N_E by
the proper density N'_E and uses the special relativistic expression (A.5) for z. This confirms a remark to this effect
in Section 10.6.

A.4. VELOCITY ADDITION AND THE ABERRATION OF LIGHT

If a particle moves with constant velocity in the x - y plane
of S, its space coordinates at time t in S will be

$$x = w_x t \ (+ \text{ constant}), \ y = w_y t \ (+ \text{ constant}), \ z = 0 \quad (A.9)$$

where w_x, w_y are components of the particle velocity in S.
The constants here can always be reduced to zero by a change
of origin, and we assume this to have been done in what follows. Substitution of (A.9) into (A.2) then gives the coordinates in S' as

$$x' = \beta(w_x - v)t, \ y' = w_y t, \ z' = 0, \ t' = \beta\left(1 - \frac{vw_x}{c^2}\right)t \qquad (A.10)$$

from which one finds

$$x' = w'_x t', \ y' = w'_y t' \qquad (A.11)$$

with

$$w'_x \equiv \frac{w_x - v}{(1 - vw_x/c^2)}, \ w'_y \equiv \frac{\beta^{-1} w_y}{(1 - vw_x/c^2)}. \qquad (A.12)$$

Eqns (A.12) determine w'_x and w'_y, the velocity components of
the moving particle in S', as functions of its velocity components in S.

A similar argument holds if one considers a reference
frame S″ which is moving in the *negative* x-direction with
speed v relative to S. The Lorentz transformation for this
frame has v replaced by $(-v)$. If one also considers the particle referred to in (A.9) to move parallel to the x-axis, so
that $w_x = w$ and $w_y = 0$, the result corresponding to (A.12) is

$$w'' = \frac{w+v}{(1+vw/c^2)} \quad \text{(relativistic addition of velocities)} \quad (A.13)$$

compared to

$$w'' = w+v \qquad \text{(Newtonian addition of velocities)} \qquad (A.14)$$

for the same situation in Newtonian theory. The crucial difference is that in special relativity the result of adding two
velocities less than c is itself less than c; for instance

$w = v = 0.8c$ leads to $w'' = 1.6c$ (Newtonian) but to $w'' = 0.9756c$ (relativistic).

To discuss aberration we return to (A.12) and suppose that the particle considered is a photon with a velocity making an angle ϑ with the negative x-direction in S, and an angle ϑ' with the negative x-direction in S'. Since the speed of light is c in *both* reference frames, we have

$$w_x = -c \cos \vartheta, \quad w_y = c \sin \vartheta \qquad (A.15)$$

and

$$w'_x = -c \cos \vartheta', \quad w'_y = c \sin \vartheta' \qquad (A.16)$$

Using (A.15) in conjunction with (A.12) gives

$$w'_x = - \frac{(c \cos \vartheta + v)}{(1 + v \cos \vartheta/c)}, \quad w'_y = \frac{c \sin \vartheta}{\beta(1 + v \cos \vartheta/c)} \qquad (A.17)$$

and it is now possible to verify that the speed of the photon in S' is c, by checking that $(w'_x)^2 + (w'_y)^2 = c^2$. Comparing (A.17) with (A.16) gives

$$\cos \vartheta' = \frac{\cos \vartheta + v/c}{1 + v \cos \vartheta/c} , \quad \sin \vartheta' = \frac{\sin \vartheta}{\beta(1 + v \cos \vartheta/c)}. \qquad (A.18)$$

If ϑ and ϑ' are small, so that $\sin \vartheta' \sim \vartheta'$, $\sin \vartheta \sim \vartheta$ and $\cos \vartheta \sim 1$, the second of these equations reduces to

$$\vartheta'/\vartheta = \frac{1}{\beta(1 + v/c)} = \left(\frac{1 - v/c}{1 + v/c}\right)^{\frac{1}{2}} \equiv \frac{1}{1 + z}. \qquad (A.19)$$

By the argument of Section 10.3 based on Fig. 10.4, but with ϑ_E replaced by ϑ', we now conclude that the energy received per unit area from a receding source must be multiplied by the aberration factor

$$f_a = (\vartheta'/\vartheta)^2 = \left(\frac{1 - v/c}{1 + v/c}\right) = \frac{1}{(1 + z)^2} \qquad (A.20)$$

Thus f_a retains the classical red-shift dependence of (10.15),

although its velocity dependence is different. From the unchanged z-dependence of f_{rs} and f_a, it follows that the luminosity distance also retains the classical form (10.21). We
have therefore confirmed remarks made to this effect in Section 10.6.

A.5. DENSITY OF GALAXIES IN THE MILNE MODEL

It is argued in Section 2.5 that the number density of a set
of galaxies, at a time t after the start of the expansion of
the universe, is in the Milne model equal to

$$N(t) = N_0\left(\frac{t_0}{t}\right)^3.$$

But if these galaxies are receding from the observer, the
question now arises in which reference frame the time t is to
be measured. If an event E occurs at a time t in the observer's frame S, and at a time t' in the frame S' which moves
with the particular galaxies under consideration, then by
(A.3) $t' = t/\beta$. (Note that this simple relationship holds for
the Milne model because each galaxy moves with constant velocity, so that there is an associated inertial frame S' in
which it is permanently at rest.) The cosmological principle
demands that the density of galaxies near any observer shall
be a function only of proper time for that observer; therefore
the relevant time is t' rather than t. We conclude that the
true density of a set of galaxies, observed by light which
left them at a time t *in* S, is

$$N(t) = N_0\left(\frac{t_0}{t'}\right)^3 = N_0\left(\frac{\beta t_0}{t}\right)^3 = N_0\left(\frac{\beta t_H}{t}\right)^3. \tag{A.21}$$

Now eqns (2.14) and (2.15) depend only on measurements in the
frame S and therefore remain valid in special relativity. It
follows from eqns (2.14) and (2.15) that

$$\frac{t_H}{t} = \frac{t_H}{r} \cdot \frac{r}{t} = \frac{v t_H}{r} = 1 + \frac{v}{c} \tag{A.22}$$

and so eqn (2.17) is to be replaced by

$$N_E = N(t) = N_0\left\{\beta\left(1+\frac{v}{c}\right)\right\}^3 = N_0\left\{\frac{1+v/c}{1-v/c}\right\}^{3/2} = N_0\left\{\frac{ct_H}{ct_H-2r}\right\}^{3/2} \quad (A.23)$$

where eqn (2.14) has again been used. This confirms eqn (10.32) of Section 10.6.

PROBLEMS

(A.1)   A spaceship leaves the Earth at $t = 0$ (in a reference
        frame S fixed in the Earth) and moves with constant speed
        $v = 0.6c$ relative to S for a period of 4 years as mea-
        sured by the spaceship's clock. It then reverses its
        direction of motion instantaneously, keeping the speed
        unchanged, and returns to Earth. During its flight it
        emits radio signals at the end of every year of flight,
        as measured by its clock, which are received by an obser-
        ver on the Earth. Assuming that (A.3) is valid during
        both parts of the flight,
        (i)   Find the ship's x and t coordinates in reference
               frame S at the time of emission of each signal.
        (ii)   Find the time of observation of each signal in S.
               Verify that these times are consistent with
               $\Delta t_0/\Delta t_E = 1+z$, if z is given by (A.6) with
               $v = +0.6c$ for signals emitted during the first
               part of the flight, and $v = -0.6c$ for signals
               emitted during the second part.
        (iii) Verify that during the flight ten years have
               elapsed on the Earth, while only eight years have
               elapsed on the ship. This is an instance of the
               famous 'clock paradox'.
        (The time difference is greater for higher speeds. If
        $v = 0.995c$, the elapsed time for a return trip to a star
        40 light-years distant is about 80 years on Earth but less
        than 8 years on the ship.)

(A.2)   A light source G is moving with speed v in S, and its
        direction of motion makes an angle ϑ with OG, O being an

observer fixed in S. Show that eqn (10.4) must be re-placed by

$$\Delta t_0 = \Delta t_E\left(1+\frac{v\ \cos\ \vartheta}{c}\right)$$

and that using (A.3) and (A.4) gives for the red-shift

$$1 + z = \beta\left(1+\frac{v\ \cos\ \vartheta}{c}\right).$$

Verify that for $\vartheta = 0$ this reduces to (A.6), and that for $\vartheta = 180°$ it reduces to (A.6) with v replaced by $(-v)$. Show also that for a given speed v there is just one value of ϑ which results in zero red-shift, and find this value for $v = 0.8c$.

(A.3) It has been suggested that some quasars are objects ejected at high speed from nearby galaxies. Suppose that such objects, all having the same intrinsic brightness, are ejected in different directions but with the same speed from a galaxy at rest relative to our own. If the maximum red-shift observed is $z = 2$, find

(i)   the speed of ejection,

(ii) the maximum blue-shift, i.e. the largest *negative* value of z,

(iii) the ratio of the maximum and minimum apparent brightnesses, assuming all the objects to be at essentially the same distances from the observer.

(It may be assumed that all special relativistic results of Tables 10.1 and 10.2 remain valid for a light source approaching the observer, with v replaced by $(-v)$.)

APPENDIX B

The mass of a typical stable particle, the number of such particles in the universe, and cosmological coincidences

A time-dependent gravitational constant has been envisaged in Section 1.2, and some further implications of it are pursued in this Appendix, where an attempt is made to relate the typical mass of a stable elementary particle to cosmological quantities [1]. If such an attempt were successful in quantitative detail, it would represent a profound advance in our understanding of both the structure of matter and the nature of the universe. In fact only rough estimates are possible and they will be made on the basis of a clear set of assumptions:

(i)   Theories are considered which involve the fundamental parameters: Planck's constant ($\hbar$), the velocity of light *in vacuo* (c), the gravitational constant (G) and the Hubble parameter (H). Their numerical values are assumed known.

(ii) H and G depend on time, but their ratio is independent of time. This assumption is sometimes called the Dirac hypothesis, which he has taken up recently in an amended form [2].

(iii) Particles whose bare rest mass is time independent will be identified with the main stable particles.

A remark is in order on the origin of assumption (ii). This arises from the following idea. The ratio of the electrostatic to the gravitational forces between proton and electron is a huge number

$$\frac{e^2/r^2}{Gm_p m_e/r^2} \sim 10^{40}.$$

Let the age of the universe — or more precisely the time since the last big-bang — be t. From the reciprocal of the present value of the Hubble constant it is approximately

$$t \sim H_0^{-1} \sim 10^{17} \text{ s}.$$

One then finds that t expressed in terms of an atomic unit of time such as $e^2/m_e c^3 \sim 10^{-23}$ s is also of order 10^{40}, m_e being the electron rest mass. An empirical relation results, of the type (other units of time and other masses can be chosen)

$$G \sim \frac{e^4}{tm_e^2 m_p c^3} \cdot$$

Now the rough equality of two distinct huge dimensionless numbers is perhaps no accident, but could be fundamental, and therefore valid at all times. This 'large number hypothesis' implies a time-dependence of G such that

$$G(t) \propto H(t),$$

as assumed in (ii).

We now start our argument by assuming that if α, β, γ, δ be unknown constants, then a typical mass in such a theory has the form

$$[\hbar^\alpha H^\beta G^\gamma c^\delta] = [ML^2 T^{-1}]^\alpha [T]^{-\beta} [M^{-1} L^3 T^{-2}]^\gamma [LT^{-1}]^\delta$$

$$= [M]^{\alpha-\gamma} [L]^{2\alpha+3\gamma+\delta} [T]^{-\alpha-\beta-2\gamma-\delta}$$

where the dimensions of each factor have been written down. The expression has to be of the dimensional form $[M]$. The exponent of $[L]$ is zero and the exponent of $[M]$ is unity if

$$2(\gamma+1) + 3\gamma + \delta = 0 \quad \text{i.e. } 5\gamma = -\delta-2$$

so that

$$\alpha = 1 + \gamma = 1 - \frac{\delta}{5} - \frac{2}{5} = \frac{3}{5} - \frac{\delta}{5}$$

The exponent of T is zero if

$$-\delta = \alpha + \beta + 2\gamma = \beta + (1+\gamma) + 2\gamma = \beta + 3\gamma + 1 = \beta + 1 - \frac{3}{5}(2+\delta).$$

It follows that

$$\beta = \frac{1}{5} - \frac{2}{5}\delta.$$

Combining these results

$$[M]^5 = \hbar^{3-\delta}H^{1-2\delta}G^{-\delta-2}c^{5\delta} = \frac{\hbar^3 H}{G^2}\left(\frac{c^5}{\hbar H^2 G}\right)^{\delta}.$$

The masses which can occur in this theory are seen to be of the form

$$m(b) = k(b)\left[\frac{\hbar^3 H}{G^2}\right]^{1/5}\left(\frac{c^5}{\hbar H^2 G}\right)^{b/15} \tag{B.1}$$

where $b = 3\delta$ is an unidentified constant and $k(b)$ is a dimensionless multiplier. In order to determine the time dependence of $m(b)$, one can re-express the H-dependences in terms of a dependence on H/G, which is independent of time. One is then left with a time dependence given by

$$m(b) \propto G^{-(b+1)/5}. \tag{B.2}$$

It follows from assumption (iii) that the order of magnitude of the stable particles is given by $b = -1$. Treating the constands $k(b)$ as of order unity, one finds

$$m \sim m(-1) \sim \left(\frac{\hbar^3 H}{Gc}\right)^{1/3} \sim 10^{-25} \text{ g.} \tag{B.3}$$

Taking the present value of H^{-1} as 5.7×10^{17} s, $\hbar = 1.05 \times 10^{-34}$ J s, $G = 6.67 \times 10^{-11}$ Nm^2 kg^{-2}, $c = 3.00 \times 10^8$ m s^{-1}, one finds (Fig. B.1)

$$\log_{10} m(b) = -16.87 + 8.14b. \tag{B.4}$$

The electron mass (9.11×10^{-31} kg), the proton mass (1.67×10^{-27} kg), and the mass of the Ω^- particle (2.98×10^{-27} kg) are also shown on the curve. It is seen that the stable particle rest masses do indeed cluster around the value $b = -1$.

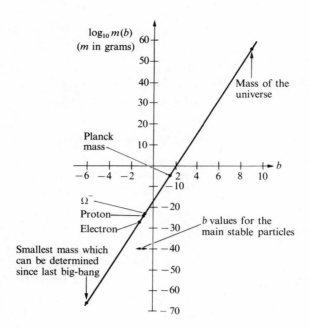

FIG. B.1 The main masses in the universe according to eqn (B.1) with $k(b) \sim 1$. Eqn (B.4) gives the equation for the straight line.

Using the critical density (6.1)

$$\rho \sim \rho_c = 3H^2/8\pi G(t),\qquad (B.5)$$

and taking the radius of the observable universe to be a horizon distance of the type introduced in eqn (10.30),

$$r_h \sim ct_H = c/H,\qquad (B.6)$$

the mass of the observable universe is of order

$$M \sim \frac{4\pi}{3}\,\rho_c r_h^3 = \frac{c^3}{2HG} = \tfrac{1}{2}m(9)\qquad (B.7)$$

This mass is also shown on the curve and is of order $10^{56.4}$. The equivalent number of stable non-interacting particles in the universe is

$$N \sim \frac{m(9)}{m(-1)} = \left(\frac{c^5}{GH^2\hbar}\right)^{2/3} \sim 2.4 \times 10^{81}. \qquad \text{(B.8)}$$

This was first estimated by A.S. Eddington in 1944 by methods not now accepted as meaningful.

We next consider the ratio Φ of the gravitational to the electric forces between two particles, using the famous fine-structure constant

$$\alpha \equiv e^2/\hbar c \sim 1/137.$$

One finds, using eqn (B.3) for the value of Φ,

$$\Phi = \frac{Gm^2/r}{e^2/r} = \frac{1}{\alpha} \frac{G[m(-1)]^2}{\hbar c} = \frac{1}{\alpha}\left(\frac{\hbar H^2 G}{c^5}\right)^{1/3} = \frac{1}{\alpha\sqrt{(N)}} \sim 2.8 \times 10^{-39} \quad \text{(B.9)}$$

This ratio is seen to be very small. One can say, therefore, on the assumption of a decreasing value of G, that the ratio Φ is so small because so much time has elapsed since the last big bang. An estimate for the decrease of G with time has been given in eqn (6.14), and the general problem has been discussed at length in the literature [3,4].

Another important consequence of assumption (ii) is that it effects a connection for simple cosmological models between the Hubble time t_H, the deceleration parameter and $\dot{G}/G$. For since $G \propto H$, one has from eqn (4.28)

$$1 + q_0 = -(\dot{H}/H^2)_{t_0} = -(\dot{G}/G)_{t_0}\, t_H.$$

Using eqn (6.14), one finds

$$q_0 \sim (8 \times 10^{-11})(1.8 \times 10^{10}) - 1 = 0.44$$

in good agreement with observational estimates (Section 11.5). The analogue of the quantity $\alpha = e^2/\hbar c$ is

$$\alpha_G \equiv \frac{Gm^2}{\hbar c} \sim \frac{G[m(-1)]^2}{\hbar c} \sim \frac{1}{N^{1/2}} \sim 2 \times 10^{-41}. \qquad \text{(B.10)}$$

This is the gravitational 'fine-structure constant'. It is

the gravitational analogue of α with the 'coupling parameter', which governs the interaction, Gm^2 instead of e^2. In the spectrum of hydrogen-like atoms of nuclear charge $+ Z|e|$, the nth principal energy level is given by quantum mechanics to be approximately

$$E_n = \frac{E_0 Z^2}{n^2} + \frac{E_0 \alpha^2}{n^4} \left(\frac{n}{k} - \frac{3}{4} \right) \tag{B.11}$$

where n is the principal quantum number ($n = 1, 2, \ldots$), k is the azimuthal quantum number,

$$E_0 \equiv \frac{me^4}{2\hbar^2} \sim 13.6 \text{ eV}, \tag{B.12}$$

and m is the rest mass of the electron. Thus α affects the spectrum, but only its finer details because of the smallness of α.

It is possible to imagine a gravitational atom bound by gravitational rather than Coulombic forces and then α_G takes the place of α in the theory of the spectra of such atoms. However, this is merely an instructive exercise since such atoms have not been found. In fact the 'radius' of the ground state $n = 1$ of an electron in a hydrogen atom is

$$\frac{\hbar^2}{\mu e^2} \sim 0.53 \times 10^{-10} \text{ m}$$

where

$$\mu = \frac{m_p m}{m_p + m}$$

is the reduced mass of a hydrogen atom arising from the motion of the electron and the proton. In fact $\mu \sim m$. The corresponding radius of a gravitational hydrogen atom is

$$\frac{\hbar^2}{\mu_G Gm_n^2} = \frac{2\hbar^2}{Gm_n^3} \sim 7.2 \times 10^{22} \text{ m} \tag{B.13}$$

where μ_G is the reduced mass for the motion of two neutrons of mass m_n each:

$$\mu_G = \frac{m_n^2}{2m_n} = \frac{1}{2} m_n.$$

This radius is only somewhat smaller than the radius of the observable universe

$$\frac{c}{H_0} \sim (3 \times 10^8) \times (5.68 \times 10^{17}) \sim 17 \times 10^{26} \text{ m}.$$

It is not expected that one will often come across dimensionless numbers which are very large, such as N or $N^{\frac{1}{2}}$, or very small such as N^{-1} or $N^{-\frac{1}{2}}$. The value of dimensional analysis in fact depends crucially on this fact.

Thus the analysis presented so far has assumed that the constant multipliers $k(b)$ in eqn (B.1) are of order unity. None the less one finds pairs of dimensionless numbers which are both very large or both very small. For example, the radius of the visible universe is by eqns (B.6) and (B.7) of order

$$\frac{c}{H} \sim \frac{GM}{c^2},$$

while the Compton wavelength of a stable particle is

$$\hbar/mc.$$

The ratio is the dimensionless number already met in eqn (B.10):

$$GMm/\hbar c \sim 10^{40} \sim N^{\frac{1}{2}}. \tag{B.14}$$

Such rough identities between very large or very small numbers are called cosmological coincidences and eqn (B.14) represents one of these. Their surprising nature must be removed by the present approach since one might otherwise take the multipliers $k(b)$ to be other than of order unity.

The coincidence eqn (B.14) will now be explained. It is an algebraic consequence of eqn (B.1) that

$$\frac{m(9)}{m(-1)} = \left[\frac{m(3/2)}{m(-1)}\right]^4. \tag{B.15}$$

The left-hand side of eqn (B.14) is

$$\frac{Gm(9)m(-1)}{\hbar c} = \frac{m(9)m(-1)}{[m(3/2)]^2} = \frac{m(9)}{m(-1)} \cdot \left[\frac{m(-1)}{m(3/2)}\right]^2. \tag{B.16}$$

By eqns (B.15), (B.16), and (B.8) the left-hand side of eqn (B.14) is

$$\left[\frac{m(9)}{m(-1)}\right]^{\frac{1}{2}} \sim N^{\frac{1}{2}}$$

as was to be established. Other identities can be obtained similarly. The mass $m(3/2)$ is called the Planck mass.

PROBLEMS

(B.1) Verify the expressions for $m(b)$ given below

b	9	3/2	-1	-6
$m(b)$	c^3/GH	$(\hbar c/G)^{\frac{1}{2}}$	$(\hbar^2 H/Gc)^{1/3}$	$\hbar H/c$

(B.2) Show from the uncertainty relation $\Delta E \Delta t \gtrsim \hbar$ that the smallest mass difference which can be measured during the period since the last big bang is of order $m(-6)$ [5].

(B.3) If the masses can depend on the parameters $\hbar$, H, G, c, and also the electronic electric charge, show from dimensional analysis that the masses have the form

$$m(b,b') \sim \left(\frac{\hbar^3 H}{G^2}\right)^{1/5} \left(\frac{c^5}{\hbar H^2 G}\right)^{\frac{b}{15}+\frac{b'}{5}} \alpha^{b'}$$

where α is the fine-structure constant. Hence show that the time-independent masses have the form

$$\alpha^{b'} m(-1)$$

where $m(-1)$ is the mass introduced in the text, and b' is an unidentified constant.

(B.4)   In a theory with electric coupling, the parameters $\hbar$, H, c occur.   Show from dimensional analysis that there is only one possible combination of constants which is dimensionally an electric charge, and that this is time independent.

(B.5)   'If one expresses the age of the universe in terms of a unit of time provided by atomic constants, say e^2/mc^3, one gets a large dimensionless number t, which is somewhere around 10^{39}' [2].

   Show that the theory of the text gives

$$t \sim N^{\frac{1}{2}}/\alpha.$$

which does yield a large $t \sim 10^{42}$.

(B.6)   Show that the gravitational radius r_G of a stable particle in analogy with the classical radius $r = e^2/\mu c^2 \sim 2.8 \times 10^{-15}$ m, satisfies

$$\frac{r_G}{r_h} \sim \frac{1}{N}$$

where r_h is given in eqn (B.6).

(B.7)   The theory outlined implies that $m(b)$ increases, remains constant, or decreases with time, depending on whether $b > -1$, $b = -1$, or $b < -1$.   The number (B.8) of stable particles in the universe was therefore unity at some time in the past.   Obtain an expression for the Hubble time t_p for this condition and verify that it satisfies an uncertainty relation with the Planck energy $m(3/2)c^2 \equiv E_p$:

$$t_p \sim \left[\hbar G(t_p)/c^5\right]^{\frac{1}{2}}, \; E_p t_p \sim \hbar$$

If the present value of G were used, verify that $t_p \sim 10^{-43}$s. [Cosmology as described must break down for $t < t_p$. The time t_p is called the Planck time.]

APPENDIX C

Representation of cosmological models in the density parameter-deceleration parameter plane

C.1. SOME CURVES IN THE (σ, q) PLANE

One way of seeing the relationship between the zero-pressure Friedmann models is to represent them in various regions of the deceleration parameter (q)-density parameter (σ) plane (Fig. C.1.). To achieve this, we first prove a preliminary result. Recall that

$$(5.8): \qquad\qquad C \equiv \frac{8\pi}{3} G \rho R^3. \qquad\qquad (C.1)$$

$$(4.25): \qquad\qquad \sigma \equiv \frac{4\pi G\rho}{3H^2} = \frac{C}{2H^2 R^3}. \qquad\qquad (C.2)$$

$$(4.24, 4.27): \quad L \equiv \frac{\lambda}{3H^2} = \sigma - q, \quad K \equiv \frac{kc^2}{R^2 H^2} = 3\sigma - q - 1. \qquad (C.3,4)$$

$$(7.1): \qquad\qquad \lambda^* \equiv \frac{4c^6}{9C^2} \qquad\qquad (C.5)$$

It then follows that by $k^3 = k$, (C.5), (C.2), (C.3), (C.4), taken in that order,

$$\frac{\lambda}{k\lambda^*} = \frac{\lambda}{k^3\lambda^*} = \frac{9\lambda C^2}{4c^6 k^3} = \frac{9\lambda}{4c^6 k^3} 4\sigma^2 H^4 R^6 = \frac{9\sigma^2 H^4 R^6}{k^3 c^6} 3H^2(\sigma-q)$$

$$= \frac{1}{k^3}\left(\frac{HR}{c}\right)^6 27\sigma^2(\sigma-q) = 27 \frac{\sigma^2(\sigma-q)}{(3\sigma-q-1)^3}. \qquad (C.6)$$

Multiplying out,

$$(27\sigma^3 k - 27\sigma^2 qk) \frac{\lambda^*}{\lambda} = 27\sigma^3 - 27\sigma^2(q+1) + 9\sigma(q+1)^2 - (q+1)^3.$$

Rearranging,

$$27\sigma^3\left(\frac{k\lambda^*}{\lambda}-1\right) + 27\sigma^2 q\left(1-\frac{k\lambda^*}{\lambda}\right) + 27\sigma^2 - 9\sigma(q+1)^2 + (q+1)^3 = 0.$$

This may be written as

$$27\sigma^2(\sigma-q)\left(\frac{k\lambda^*}{\lambda}-1\right) + G(\sigma,q+1) = 0 \qquad (C.7)$$

where

$$G(\sigma,x) \equiv 27\sigma^2 - 9\sigma x^2 + x^3. \qquad (C.8)$$

The significance of (C.6) or (C.7) is that, given the value of $k\lambda^*/\lambda$, it assigns to each q a corresponding σ. Alternatively, given a point in the (σ,q)-plane (C.7) determines the corresponding value of $k\lambda^*/\lambda$.

As an example of the use of these relations, let us find the equation of the curve $\lambda = \lambda^*$ in the $k = +1$ region on the q-σ diagram. By (C.7) we see at once that the required equation is $G(\sigma,q+1) = 0$. This is a quadratic equation in σ with solutions

$$\sigma_{1,2} = \frac{1}{3}\left\{ \frac{(q+1)^2}{2} \pm \sqrt{\left(\frac{(q+1)^4}{4} - \frac{(q+1)^3}{3}\right)} \right\} \qquad (C.9)$$

For a real root we need $q \geqslant 1/3$ which leads to $\sigma_1 \geqslant 16/54 = 0.296$; but this is not permissible as it would lead by (A.3) to $\lambda^* < 0$. The least possible value of q for $\lambda = \lambda^*$ is actually $q = \frac{1}{2}$ when

$$\sigma_{1,2} = \frac{1}{3}\left(\frac{9}{8} \pm \frac{3}{8}\right) \text{ i.e. } \sigma_1 = \frac{1}{2}, \ \sigma_2 = \frac{1}{4}.$$

σ_2 is always to be rejected as it is less than q and implies $\lambda^* < 0$. Hence

$$\lambda = \lambda^* \text{ implies } \sigma = \frac{1}{3}\left\{\frac{(q+1)^2}{3} + \sqrt{\left(\frac{(q+1)^4}{4} - \frac{(q+1)^3}{3}\right)}\right\} \text{ and } \sigma,q \geqslant \frac{1}{2}.$$
$$(C.10)$$

This curve is shown in Fig. C.1, and divides regions 4a and 4b [1,2].

The next curve to be obtained is of the form $q \geqslant F(\sigma)$, where $F(\sigma)$ is the least permitted value of q if σ is given. If $q < -1$ then $k > 0$, and hence $k = +1$ by (C.4). Now for $k = +1$ and $\lambda < \lambda^*$ one has (by Fig. 7.2) oscillating models and we know that $q > 0$ for these [see (7.16]. It follows that

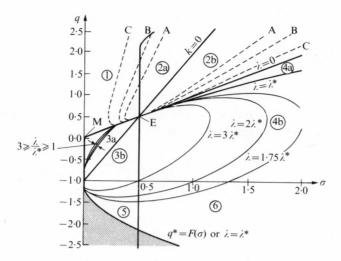

FIG. C.1. Deceleration—density parameter relations for various cosmological constants. M and E denote respectively the Milne and the Einstein-de Sitter models. A: Curve for $\lambda = -15\lambda*$, B: curve for $\lambda = -5\lambda*$, C: curve for $\lambda = -\lambda*$.

$$q < -1 \text{ implies } k = +1 \text{ and } \lambda > \lambda*. \tag{C.11}$$

It also follows from (C.3) that $\sigma > q$. The first expression in (C.7) is therefore negative and one can now find from (C.7) that

$$q < -1 \text{ implies } G(\sigma, q+1) > 0. \tag{C.12}$$

This is the required constraint. Whereas in (C.9) we solved $G(\sigma, q+1) = 0$ for σ, which yielded a quadratic equation, we imagine in (C.12) that σ is given and obtain an inequality for q. This yields a cubic *equation* by virtue of (C.8), as will now be shown. Since

$$\left[\frac{\partial G(\sigma, q+1)}{\partial(q+1)} \right]_{\sigma} = (q+1)[3(q+1)-18\sigma] = 3(q+1)(q+1-6\sigma)$$

is positive for $q+1 < 0$, $G(\sigma, q+1)$ increases with q. Hence we can first solve the cubic $G(\sigma, q+1) = 0$ for q. Let the solution be $q^* = F(\sigma)$. The condition $G(\sigma, q+1) > 0$ is then fulfilled if

$$q > q^* = F(\sigma) \qquad (q < -1). \qquad \text{(C.13)}$$

In order to solve the cubic, let $y \equiv q + 1 - 3\sigma$. Then the equation for y is a 'reduced' cubic equation, i.e. a cubic equation without the quadratic term:

$$G(\sigma, q+1) \equiv y^3 + 3py + 2t = 0,$$

$$p \equiv -(3\sigma)^2, \qquad t \equiv \frac{27}{2} \sigma^2 (1-2\sigma).$$

The nature of the roots of such an equation is determined by the values of

$$p^3 + t^2 = \frac{3^6}{4} \sigma^2 (1-4\sigma).$$

The discussion of the various solutions is routine, and is not given here (see Problem (C.1)). It leads to

$$F(\sigma) = \begin{cases} 3\sigma\{1-2 \cos \frac{\varphi-\pi}{3}\} - 1 & (\sigma > 1/4) \\[2ex] 3\sigma\left\{1 + \left[1 - \frac{1}{2\sigma} + \sqrt{\left(\frac{1}{4\sigma^2} - \frac{1}{\sigma}\right)}\right]^{1/3} + \left[1 - \frac{1}{2\sigma} - \sqrt{\left(\frac{1}{4\sigma^2} - \frac{1}{\sigma}\right)}\right]^{1/3}\right\} & (\sigma < 1/4) \\[2ex] -7/4 & (\sigma = 1/4) \end{cases}$$

$$\text{(C.14)}$$

where φ is defined by $\cos \varphi = 1 - 1/2\sigma$. This curve is also shown in Fig. C.1 as the lower limit of the q-values for regions 5 and 6 [1].

The only other curve requiring some calculations is that separating regions 1 and 2a. We turn to it next. Using the eqn (4.15)

$$\dot{R}^2 = \frac{C}{R} + \frac{1}{3}\lambda R^2 - kc^2,$$

we substitute in it from (C.2), (C.3), and (C.4) for C, λ, and

$k\sigma^2$, using values for some arbitrary but fixed time t_1, to find

$$\dot{R}^2 = (R_1 H_1)^2 \left[2\sigma_1 \frac{R_1}{R} + (\sigma_1 - q_1)\left(\frac{R}{R_1}\right)^2 - 3\sigma_1 + q_1 + 1 \right].$$

Let $y = R/R_1$ and write the relation as

$$\dot{y}^2 = H_1^2 f(y), \quad f(y) \equiv \frac{2\sigma_1}{y} + (\sigma_1 - q_1)y^2 - 3\sigma_1 + q_1 + 1.$$

It follows that if R is taken as zero when $t = 0$, then $\dot{y} = H_1 \sqrt{f(y)}$ can be integrated to yield [3]

$$H_1 t_1 = \int_0^1 dy / \sqrt{(f(y))}. \qquad (C.15)$$

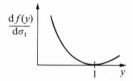

FIG. C.2. A simple property of $f(y)$.

The integrand increases ($f(y)$ decreases) as σ_1 increases in the range of integration (see Fig. C.2). The lowest value of σ_1, $\sigma_1 = 0$, therefore furnishes the inequality for $q > 0$

$$Ht < \int_0^1 \frac{dy}{\sqrt{\left(q + 1 - qy^2\right)}} = \frac{1}{\sqrt{q}} \int_0^1 \frac{dy}{\sqrt{\left(\left(1 + \frac{1}{q}\right) - y^2\right)}} = \frac{1}{\sqrt{q}} \sin^{-1}\left[\frac{y\sqrt{q}}{\sqrt{(1+q)}}\right]_0^1$$

where the suffixes 1 have again been omitted. Hence [4]

$$Ht < h(q) \equiv \begin{cases} \frac{1}{\sqrt{q}} \sin^{-1} \sqrt{\left(\frac{q}{1+q}\right)} & (q > 0) \\[2ex] \frac{1}{\sqrt{(-q)}} \sinh^{-1} \sqrt{\left(\frac{-q}{1+q}\right)} & (0 > q > -1). \end{cases} \qquad (C.16)$$

We have used the opportunity to define a function $h(q)$ and to give the integral for $0 > q > -1$. For $q < -1$, $h(q)$ is imaginary and the inequality fails. The form of $h(q)$ is shown in Fig. C.3.

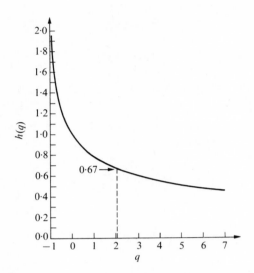

FIG. C.3. The function $h(q)$ which represents an upper limit on the present age of the universe in units of H_0^{-1}. The curve is a plot of eqn (C.16) and the arrow marks the limit (C.26).

The inequality (C.16) can actually be improved by noting that, dropping suffices 1,

$$f(y) \gtrless \frac{2\sigma}{y} + (\sigma-q) - 3\sigma + q + 1 \quad \text{for } q \gtrless \sigma \quad (C.17)$$

in the range $1 > y > 0$. Integration as in eqn (C.15), but using the right-hand side in the integrand, now yields (see Problem (C.3))

$$Ht \gtrless j(\sigma) \quad \text{for } \sigma \gtrless q \quad (\sigma > \tfrac{1}{2}) \quad (C.18)$$

where the signs are to be taken in order and

$$j(\sigma) \equiv \frac{2\sigma}{(2\sigma-1)^{3/2}} \left[\frac{\pi}{2} - \sin^{-1} \frac{1}{\sqrt{(2\sigma)}} \right] - \frac{1}{2\sigma-1}. \quad (C.19)$$

The equation $h(q) = j(\sigma)$ for $q > 0$, $\sigma > \tfrac{1}{2}$ forms part of the boundary between regions 1 and 2 in Fig. C.1.

C.2.   THE REGIONS OF THE (σ, q)-HALF PLANE (Fig. C.1)

Each big-bang Friedmann model can be represented as a point on
the $(\sigma\text{-}q)$-half plane. However, the region $q < F(\sigma)$ is not
allowed by eqn (C.13) and is stippled in Fig. C.1 to indicate
this. The permissible part of the plane can be divided into
regions which are discussed individually below. The properties of these regions are summarized in Table C.1.

Region 1 (defined by the curve $h(q) = j(\sigma)$ and the lines
$2\sigma = 1$ and $\lambda = 0$).

   Result (C.18) holds in the part of the region in which
$2\sigma > 1$, but, in this part, $h(q) < j(\sigma)$ so that result (C.18)
is weaker than (C.16). From eqns (C.3) and (C.16) one can
put, since $q > -1$ throughout the region,

$$Ht = \frac{\lambda t}{3(\sigma - q)H} < h(q), \qquad (C.20)$$

which is an inequality for λ (see Table C.1). Similarly,
from eqns (C.4) and (C.16)

$$Ht = \frac{c^2 kt}{(3\sigma - q - 1)HR^2} < h(q) \qquad (C.21)$$

since $q > \sigma$, $\lambda < 0$ by eqn (C.3). Also, since eqn (C.4) represents a straight line of slope 3 from the point $(0, -1)$ in the
(σ, q)-half plane at each instant, then $k = -1$ in the lower
part of region 1. In fact, computation shows that $k = -1$ for
$0 < q < 13.2$.

Region 2 (defined by different parts of the curves used for
region 1).

   Result (C.18) is now stronger than inequality (C.16) and
the region has been designed so that eqn (C.18) applies
throughout it. Since $q > \sigma$, $\lambda < 0$ as in region 1. The $k = 0$
line, to the left of which $k = -1$ and to the right of which
$k = +1$, crosses this region. The inequalities for λ and
kR^{-2} are given by eqns (C.20) and (C.21) with $j(\sigma)$ instead of

TABLE C.1

Algebraic inequalities for various regions of the (σ,q)-half plane

$$L = \lambda/3H^2, \quad K = kc^2/H^2R^2$$

Number of region	Result used	λ-inequality	k-inequality	Additional comments
1	(C.16) $Ht < h(q)$	$L < \dfrac{(\sigma-q)h(q)}{Ht} < 0$	$K < \dfrac{(3\sigma-q-1)h(q)}{Ht}$	$\lambda < 0,\ q > \sigma,\ k = -1$
2	(C.18) $Ht < j(\sigma)$	$L < \dfrac{(\sigma-q)j(\sigma)}{Ht} < 0$	$K < \dfrac{(3\sigma-q-1)j(\sigma)}{Ht}$	$\lambda < 0,\ q > \sigma;\ k = -1$ for $3\sigma < q+1$; $k = 1$ for $3\sigma > q+1$
3	(C.16) $Ht < h(q)$	$0 < L < \dfrac{(\sigma-q)h(q)}{Ht}$	$K < \dfrac{(3\sigma-q-1)h(q)}{Ht}$	$\lambda > 0,\ \sigma > q;\ k = -1$ for $3\sigma < q+1$; $k = 1$ for $3\sigma > q+1$
4	(C.16, C.18) $j(\sigma) < Ht < h(q)$	$j(\sigma) < L < \dfrac{Ht}{\sigma-q} < h(q)$	$j(\sigma) < K\dfrac{Ht}{3\sigma-q-1} < h(q)$	$\lambda > 0,\ \sigma > q,\ k = 1$
5	(C.13) $q > F(\sigma)$	$0 < L < \sigma - F(\sigma)$	$0 < K < 3\sigma-1-F(\sigma)$	$\sigma > 0 > q,\ k = 1,\ -1 > q > -2.1$
6	(C.13, C.18) $q > F(\sigma),\ Ht > j(\sigma)$	$\dfrac{(\sigma-q)j(\sigma)}{Ht} < L < \sigma-F(\sigma)$	$\dfrac{(3\sigma-q-1)j(\sigma)}{Ht} < K < 3\sigma-1-F(\sigma)$	$\sigma > 0 > q,\ k = 1,\ -2.1 > q$

For $k = +1$ and $\lambda < \lambda^*$ one has a closed big-bang universe which cannot expand without limit. This is a 'black-hole' type universe [5].

$h(q)$. The curve $3\sigma = q + 1$ or $k = 0$ divides the region into two parts 2a and 2b.

Region 3 (defined by the straight lines $\lambda = 0$, $2\sigma = 1$, $q = -1$).
        Neither result (C.13) nor result (C.18) applies, but (C.16) can be used. It changes from the sin to the sinh form for $q < 0$. Since $\sigma > q$, $\lambda > 0$ by (C.3), k changes sign in the region as shown in the diagram, dividing it in regions 3a and 3b. A $\lambda = \lambda^*$ curve exists in region 3a, but in region 3b the smallest λ value is $3\lambda^*$ and it occurs at the corner $q = -1$, $\sigma = \frac{1}{2}$, as can be verified from eqn (C.7).

Region 4 (defined by different parts of the lines used for region 3).
        (C.16) and (C.18) both apply and yield upper and lower bounds respectively for Ht, λ, and k/R^2 in a manner analogous to that explained for region 1. The curve $\lambda = \lambda^*$ given by (C.10) divides this region into parts a and b.

Region 5 (defined by the curve $q = F(\sigma)$ and the straight lines $2\sigma = 1$ and $q = -1$).
        Result (C.15) applies in this region. Also from (C.3) and (C.4)

$$q = \sigma - \lambda/3H^2 > F(\sigma) \tag{C.22}$$

and

$$q = 3\sigma - 1 - kc^2/H^2R^2 > F(\sigma) \tag{C.23}$$

which give inequalities for λ and for k/R^2. Since $\sigma > q$ one has $\lambda > 0$ by (C.3) and, since the region lies below the dashed line of $k = 0$ (See Fig. C.1), one must have $k = +1$.

Region 6 (defined by different parts of the curves used for region 5).
        Both results (C.13) and (C.18) apply. Inequalities (C.20) and (C.21) hold with h replaced by j and the inequali-

ties reversed. In addition, inequalities (C.22) and (C.23) hold.

C.3. SOME NUMERICAL ESTIMATES

Let us now assume that the upper bound for the present value of the density parameter is its value for dense clusters [6]

$$0 < \sigma_0 \leqslant 28.4. \tag{C.24}$$

For $x \equiv q_0 + 1 < 0$ we have $G(\sigma_0, x) > 0$ by (C.12), and this implies $\sigma > \sigma_{1,2}$ by (C.8). Using (C.9) and multiplying by 6, this inequality is

$$170.4 > 6\sigma^2 > x^2 + \sqrt{(x^4 - 4x^3/3)}.$$

This leads to $-9.06 < x < 0$.   Hence [4]

$$0 < \sigma_0 \lesssim 28.4 \text{ and } q_0 < -1 \text{ implies } -10.1 < q_0 \tag{C.25}$$

Next, assume that at the present time

$$H_0^{-1} < 18 \times 10^9 \text{ years}, \ t_0 > 12 \times 10^9 \text{ years, so that } H_0 t_0 > 0.67. \tag{C.26}$$

One finds from (C.16) and Fig. C.3 that $q_0 \lesssim 2.0$.

Assumptions (C.25) and (C.22) yield

$$-10.1 < q_0 < 2.0 \tag{C.27}$$

It should be emphasized that the assumed numerical values are constantly being revised in the light of more accurate measurements, and that the limits (C.23) are therefore liable to revision. In fact, q_0 is believed to lie in a rather narrower range.

In order to obtain numerical limits on the quantities L

and K in Table C.1, note from (C.3) and (C.4)

$$L = \sigma - q, \quad K = 3\sigma - q - 1,$$

so that at the present epoch, distinguished by suffix 0,

$$\sigma_{0min} - q_{0max} < L_0 < \sigma_{0max} - q_{0min} \qquad (C.28)$$

$$3\sigma_{0min} - q_{0max} - 1 < K_0 < 3\sigma_{0max} - q_{0min} - 1. \qquad (C.29)$$

It will be assumed in addition that

$$-10.1 < q_0 < 2 \text{ and } 0 < \sigma_0 < 28.4, \qquad (C.30)$$

values which are suggested by (C.24) and (C.27). (C.28) to
(C.30) lead to Table C.2. In some cases knowledge of the
signs of K_0 and L_0 is a stronger restriction on the values of
K_0 and L_0 than is provided by (C.28) and (C.29). In such
cases the numbers furnished by (C.28) and (C.29) are placed in
brackets in Table C.2. Note that the numbers given in the
Table are independent of the present day value (H_0) of
Hubble's constant.

TABLE C.2

Numerical limits on $L \equiv \lambda/3H^2$ and $K \equiv kc^2/H^2R^2$ if (C.30) is used

Number of region	σ_{0min}	σ_{0max}	q_{0min}	q_{0max}	L_{min}	L_{max}	K_{min}	K_{max}
1	0	1.4	0	2	-2	0 (1.4)	-3.0	0 (3.2)
2	½	5	½	2	-1.5	0 (4.5)	-1.5	13.5
3	0	½	-1	½	0 (-0.5)	1.5	-1.5	1.5
4	½	28.4	-1	2	0 (-1.5)	29.4	0 (-1.5)	85.2
5	0	½	-2.1	-1	1.0	2.6	0	2.6
6	½	28.4	-10.1	-1	1.5	38.5	1.5	94.3
4a	½	11.3	½	2	0 (-1.5)	10.8	0 (-1.5)	32.4

PROBLEMS

(C.1) Establish the result (C.14).

(C.2) Show that for $\sigma \gg 1$ in (C.14)

$$\varphi \sim -1/\sqrt{\sigma}, \quad F(\sigma) \sim -\sqrt{(3\sigma)}.$$

(C.3) Establish the result (C.18).

(C.4) Show from (C.30) that

$$-2 < \frac{\lambda}{3H_0^2} < 38.5.$$

Given $H_0^{-1} < 19.4 \times 10^9$ years and that the maximum likely value of H_0 is twice its minimum value, show from the above inequality that $-6.38 \times 10^{-35}\ s^{-2} < \lambda < 12.3 \times 10^{-34}\ s^{-2}$.

(C.5) With the same assumptions show that

$$-3 < \frac{kc^2}{R_0^2 H_0^2} < 94.3$$

so that the space curvature satisfies

$$-0.319 \times 10^{-34} < \frac{kc^2}{R_0^2} < 10.0 \times 10^{-34}\ s^{-2}$$

(C.6) Show that the values $\sigma_0 = 28.4$, $H_0^{-1} \sim 10^{10}$ years correspond to a mass density of $10^{-27}\ gm\ cm^{-3} = 10^{-24}\ kg\ m^{-3}$.

(C.7) (i) Check that (C.13) for $\sigma_0 = 28.4$ yields an inequality $q_0 > -10$ in agreement with (C.25) (which is based on (C.12)).

(ii) If $\sigma_0 = 28.4$ show from (C.18) that $H_0 T_0 > 0.178$, and that this is consistent with (C.26).

(C.8) For region 5 it is stated in Table C.1 that $q > -2.1$.
Deduce this result from the definition of the shape of
the region.

(C.9) Show that for $0 < \sigma \ll 1$

$$F(\sigma) \doteqdot -1 - 3\sigma^{2/3} + 3\sigma.$$

(C.10) Show that if 1 Mpc ('one megaparsec') is 3.086×10^{24}
cm, then $H_0 = 100$ km s^{-1} Mpc^{-1} is 0.324×10^{-17} s^{-1} =
1.02×10^{-10} year^{-1} and $H_0^{-1} = 0.978 \times 10^{10}$ years.

(C.11) Suppose 40 km s^{-1} Mpc^{-1} < H_0 < 100 km s^{-1} Mpc^{-1} is a
reasonable restriction on the present value of Hubble's
constant. Show from Table C.2 that the limits on λ and
on kc^2/R_0^2, both expressed in units of 10^{-35} s^{-2}, are as
detailed in the following table.

Region Number	λ	kc^2/R_0^2
1	$-6.30 \to 0$	$-3.15 \to 0$
2	$-4.72 \to 0$	$-1.57 \to 14.2$
3	$0 \to 4.72$	$-1.57 \to 1.58$
4	$0 \to 92.6$	$0 \to 89.6$
5	$0.504 \to 8.19$	$0 \to 2.73$
6	$0.756 \to 121.0$	$0.252 \to 99.1$
4a	$0 \to 34.0$	$0 \to 34.0$

(C.12) Show that for a model universe which lies in region 4a

$$3.0 \times 10^{-27} \text{ kg m}^{-3} < \rho < 424 \times 10^{-27} \text{ kg m}^{-3}.$$

[The power of ten is -30 if cgs units are used.]

SOLUTIONS TO PROBLEMS

CHAPTER 2

(2.1) Since for a particular galaxy $r \propto R \propto t^{\alpha}$, $v = dr/dt \propto \alpha t^{\alpha-1}$ and so $H = \alpha t^{\alpha-1}/t^{\alpha} = \alpha/t$. Hence $H_0 = \alpha/t_0$ and $t_0 = \alpha/H_0 \equiv \alpha t_H$. The observed velocity − distance relation is, as in eqn (2.5),

$$v = H(t_E)r = H(t_0 - r/c)r = \frac{\alpha r}{t_0 - r/c} = \frac{\alpha r}{\alpha t_H - r/c}.$$

The density is

$$N_E = \frac{N_0}{\{\bar{R}(t)\}^3} = N_0 \left\{\frac{t_0}{t}\right\}^{3\alpha} = N_0 \left\{\frac{t_0}{t_0 - r/c}\right\}^{3\alpha} = N_0 \left\{\frac{\alpha c t_H}{\alpha c t_H - r}\right\}^{3\alpha}.$$

The Milne equations follow from $\alpha = 1$ and the Einstein−de Sitter equations from $\alpha = 2/3$.

CHAPTER 3

(3.1) Solving $r = r_1 - r_2$, $(m_1 + m_2)R = m_1 r_1 + m_2 r_2$, one finds

$$r_1 = R + \frac{m_2}{m_1 + m_2}r, \quad r_2 = R - \frac{m_1}{m_1 + m_2}r.$$

Substitution in the expression for the kinetic energy yields the required result. The first term is due to the motion of the centre of mass, and the system kinetic energy is thus decomposed into the kinetic energy of (a) bodily motion and (b) internal motion.

(3.2) (i) $\bar{E} = \bar{T} + \bar{U} = \left(\frac{s}{2} + 1\right)\bar{U}$

$\bar{E} = \left(\frac{2}{s} + 1\right)\bar{T}$

(ii)   The position vector of m relative to M is

$$\mathbf{r} = a\{\mathbf{i} \cos(\omega t) + \mathbf{j} \sin(\omega t)\}$$

where $\mathbf{i}, \mathbf{j}$ are fixed unit vectors and ω is the constant angular velocity. Differentiating gives

$$\dot{\mathbf{r}} = a\{\mathbf{j} \, \omega \cos(\omega t) - \mathbf{i} \, \omega \sin(\omega t)\},$$

$$\ddot{\mathbf{r}} = -a\omega^2\{\mathbf{i} \cos(\omega t) + \mathbf{j} \sin(\omega t)\}$$

so the speed is $v(a) = |\dot{\mathbf{r}}| = a\omega$ and the acceleration is $|\ddot{\mathbf{r}}| = a\omega^2 = v^2(a)/a$. Now $|\ddot{\mathbf{r}}|$ must equal $g(a) = GM/a^2$ (see eqn. (3.2)); therefore $v^2(a) = GM/a$ and $\bar{T} = \tfrac{1}{2}mv^2(a) = GMm/2a$. Also by eqn. (3.9) $\bar{U} = -GMm/a$ and so $\bar{E} = \bar{T} + \bar{U} = -GMm/2a$. The virial theorem holds with $s = -1$.

(3.3)   (i)   The acceleration towards the centre in a circular orbit of radius R is $(v_c(R))^2/R$, and this must equal the gravitational acceleration g, so that $v_c(R) = \surd(gR)$. The orbital period $T(R)$ is the time taken to cover the circumference $2\pi R$ at speed $v_c(R)$, so that

$$T(R) = 2\pi R/v_c(R) = 2\pi\surd(R/g).$$

The gravitational force on a mass m at distance R from the centre is equal to GMm/R^2 by eqn (3.2) and also to mg by Newton's 2nd law. It follows that $g = GM/R^2$. Substituting this into the expression for $T(R)$ gives

$$T(R) = 2\pi\surd(R^3/GM) = 2\pi\surd(3/4\pi\bar{\rho}G),$$

since $M = 4\pi\bar{\rho}R^3/3$. It follows that $T(R)$ is the same for all planets having the same mean density $\bar{\rho}$. Numerical values for the Earth are

$$v_c(R) = \sqrt{(10 \times 6.4 \times 10^6)} = 8000 \text{ m s}^{-1} = 8 \text{ km s}^{-1}$$

and

$$T(R) = 2\pi R/v_c(R) = 2\pi \times 6400 \text{ km}/8 \text{ km s}^{-1}$$

$$= 5027 \text{ s} = 83 \text{ min } 46 \text{ s}.$$

(ii) Let $g(r)$ be the gravitational acceleration at a general distance r from the centre. Then the arguments given above are valid for general r, with the replacements $R \rightarrow r$ and $g \rightarrow g(r) = GM/r^2$, and result in

$$v_c(r) = \sqrt{(g(r)r)} = \sqrt{(GM/r)} = \sqrt{(gR^2/r)} \quad (3.3.1)$$

since as established above $g = GM/R^2$. Comparison with eqn (3.21) shows that $v_e(r) = \sqrt{2}\, v_c(r)$. Also the period for distance r is $T(r) = 2\pi r/v_c(r) = 2\pi\sqrt{(r^3/GM)}$ and so by comparison with the result of part (i)

$$T(r) = T(R)(r/R)^{3/2} \quad (3.3.2)$$

as required by part (iii). Kepler's third law is the extension of this result to elliptical orbits.

The Skylab results can be obtained as follows. Comparing the expressions for $v_c(r)$ and $v_c(R)$ we find $v_c(r) = v_c(R)\sqrt{(R/r)}$. Now $R = 6400$ km, r (Skylab) $= 6800$ km and $v_c(R) = 8$ km s^{-1}, so

$$v_c(r) = 8 \ (6400/6800)^{\frac{1}{2}} \text{ km s}^{-1} = 7.76 \text{ km s}^{-1}.$$

Using eqn (3.4.2) with $T(R) = 5027$ s, we find the orbital period of Skylab as

$$T(R) = 5027 \ (6800/6400)^{3/2} \ s = 5506 \ s.$$

(iii) See proof of (3.3.2) above.

(iv) The earth rotates once in 24 hours relative to the Earth–Sun line. However, this line itself makes a complete revolution in one year — 365 days. Therefore in 24 hours the Earth–Sun line moves through 1/365 of a revolution, so the earth has made 366/365 revolutions relative to axes fixed in space. The Earth's true rotational period is therefore 365/366 × 24 hours ~ 23 hours 56 minutes ~ 86 160 s. The value of r which gives a circular orbit with this period is determined by eqn (3.3.2) and $T(R) = 5027$ s;

$$86 \ 160 = 5027 \ (r/R)^{3/2}; \quad (r/R)^{3/2} = 17.14;$$

$$r = R \ (17.14)^{2/3} = 6.648 \ R = 42 \ 547 \ km.$$

(v) It follows from eqn (3.3.2) that, for any distances r_1, r_2,

$$T(r_2) = T(r_1)(r_2/r_1)^{3/2}.$$

Taking r_1 = 9300 km (Phobos), r_2 = 23 400 km (Deimos) and $T(r_1)$ = 7 hours 42 minutes = 27 720 s, we find

$$T \ (Deimos) = 27 \ 720 \ (23 \ 400/9300)^{3/2} \ s$$

$$= 110 \ 635 \ s \sim 30 \ h \ 44 \ min.$$

Eqn (3.3.1) can be rearranged to give

$$g = r(v_c(r))^2/R^2 = 4\pi^2 r^3/R^2(T(r))^2$$

using also $T(r) = 2\pi r/v_c(r)$. Using R(Mars) =

3400 km and the Phobos data we find

$$g(\text{Mars}) = 4\pi^2 \ (9300)^3/(3400)^2(27 \ 720)^2 \ \text{km s}^{-2}$$

$$= .003575 \ \text{km s}^{-2} = 3.575 \ \text{m s}^{-2}.$$

The discrepancy between this and the entry in section 3.11 arises from the approximate values of r_1 and r_2 used here.

(vi) By eqn (3.22) the speed needed on leaving the Earth's atmosphere is

$$v(R) = \{v^2(\infty) + v_e^{\ 2}(R)\}^{\frac{1}{2}}$$

Now $v(\infty) = (\sqrt{2}-1) \ 30 = 12.43 \ \text{km s}^{-1}$, while from part (i)

$$v_e(R) = \sqrt{2} \ v_c(R) = 8\sqrt{2} \ \text{km s}^{-1}.$$

Therefore

$$v(R) = \{12.43^2 + 128\}^{\frac{1}{2}} \ \text{km s}^{-1} = 16.81 \ \text{km s}^{-1},$$

about 48 per cent higher than $v_e(R) = 11.31 \ \text{km s}^{-1}$.

(3.4) From question (3.2) one has for a circular orbit

$$\frac{1}{2} \ mv^2 = GMm/2r$$

so that at distance $R + H = r$ from the centre of the sun

$$T(r) = \frac{2\pi r}{v} = 2\pi r \sqrt{\left(\frac{r}{GM}\right)} = \frac{2\pi}{\sqrt{(GM)}} \ r^{3/2}$$

$$= \frac{2\pi R^{3/2}}{\sqrt{(GM)}}\left(1 + \frac{H}{R}\right)^{3/2},$$

H being the distance from the Sun to the earth.

This result can also be obtained from the solution given to problem 3.3 (iii)

$$T(r) = T(R)\left(\frac{r}{R}\right)^{3/2} = 2\pi \sqrt{\left(\frac{R}{g(R)}\right)}\left(\frac{r}{R}\right)^{3/2}.$$

Evaluating this numerically

$$\left(\frac{r}{R}\right)^{3/2} = \left[\frac{14.98 \times 10^{10}}{6.995 \times 10^{8}}\right]^{3/2} = (216)^{3/2} = 3182.$$

Hence

$$T(r) = \frac{2\pi \times 18.34 \times 10^{12} \times 3182}{\sqrt{(1.329)} \; 10^{10} \times 60 \times 60 \times 24} \text{ days} = 368 \text{ days}$$

(3.5) (i) $M(r) = \int_{a}^{r} \frac{k}{r} 4\pi r^2 \mathrm{d}r = 2\pi k (r^2 - a^2).$

(ii) $V(r) = - GM(r)/r$ since each shell acts as if all its mass were concentrated at the centre.

(iii) Each element of mass δm is moved from ∞ to r. At infinity its potential energy is zero. At r it is $V(r)\, \delta m$. The change in potential energy $\delta U(r) - \delta U(\infty) = \delta U(r)$ is the work done *against* the field. Summing over the elements δm which make up the thin spherical shell of mass $\mathrm{d}M(r) = (k/r)4\pi r^2 \mathrm{d}r$ yields the change in potential energy $\Sigma \delta U(r) = V(r)\, \Sigma \delta m$. We write this as $\mathrm{d}U(r) = V(r)\mathrm{d}M(r) = -\mathrm{d}W(r)$, where $\mathrm{d}W(r)$ is the work done *by* the field in producing the condensation of the thin shell.

(iv) $\int \mathrm{d}W(r) = - \int V(R)\ \mathrm{d}M(r) = \int_{a}^{b} 2\ kG\ \frac{r^2 - a^2}{r} \cdot 4\pi k r\ \mathrm{d}r$

$= \frac{8\pi^2 k^2 G}{3} (b^3 + 2a^3 - 3a^2 b).$

Now $b^3 + 2a^3 - 3a^2 b = (b-a)^2 (b+2a).$ Therefore

$$\int \mathrm{d}W(r) = \frac{8\pi k^2 G}{3} (b-a)^2 (b+2a). \ \frac{M^2}{4\pi k^2 (b^2 - a^2)^2},$$

the last factor being just unity. This yields the stated result. It is positive because force and displacement are in the same direction for the whole process.

(v)   The work done by the field is

$$U(\infty) - U(r) = \int dW(r)$$

$$\therefore U(r) = U(\infty) - \int dW(r) = - \int dW(r)$$

$$= -\frac{2}{3} GM^2 \frac{b+2a}{(b+a)^2}.$$

This is negative because gravitating masses tend to condense spontaneously from infinity, where the potential energy is zero; the later this collapse is stopped the more negative the potential energy.

(3.6) (i)   We have, on adding a layer of thickness δr, within $r + \delta r$ a mass of

$$\frac{4\pi}{3}(r+\delta r)^3 [\rho+\delta\rho] = \frac{4\pi}{3}r^3 \rho \left[1+3\frac{\delta r}{r}+\frac{\delta\rho}{\rho}\right]$$

This mass is also $m + \delta m$, i.e.

$$\frac{4\pi r^3}{3}\rho + 4\pi r^2 \sigma \, dr.$$

Equating,

$$4\pi r^2 \rho \, dr + \frac{4\pi}{3}r^3 \, d\rho = 4\pi r^2 \sigma \, dr,$$

which is the required result.

(ii)   We have $-g(r) = \frac{Gm(r)}{r^2} = \frac{G}{r^2}\cdot\frac{4\pi}{3}r^3 \rho = \frac{4\pi}{3}rG\rho$ so that

$$-\frac{dg}{dr} = -\frac{2Gm}{r^3} + \frac{G}{r^2}\frac{dm}{dr}$$

$$= -\frac{2G}{r^3}\frac{4\pi}{3}r^3 \rho + \frac{G}{r^2}\left[4\pi r^2 \rho + \frac{4\pi}{3}r^3 \frac{d\rho}{dr}\right]$$

$$= -\frac{8\pi G}{3}\rho + 4\pi G\rho + \frac{4\pi}{3}Gr\frac{d\rho}{dr}$$

$$= \frac{4\pi G}{3}\left[\rho + r\frac{d\rho}{dr}\right] = 4\pi G(\sigma - \frac{2}{3}\rho).$$

(iii)

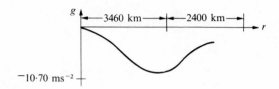

At the surface g increases with r since

$$\sigma - \frac{2}{3}\rho = 3.3 - 3.68 < 0$$

So the numerically largest value of $g(r)$ is expected *inside* the Earth (see sketch).

(3.7)   For the neutral point x

$$\frac{E}{(D-x)^2} + A = \frac{M}{x^2} + A,$$

where the masses of Earth and Moon are denoted by E and M respectively and GA is the gravitational field due to the sun.   It follows that

$$\frac{E}{D^2\left(1-\frac{x}{D}\right)^2} = \frac{M}{x^2}$$

i.e.

$$\frac{E}{MD^2}x^2 = 1 - 2\frac{x}{D} + \left(\frac{x}{D}\right)^2$$

Hence

$$\left(\frac{E}{M}-1\right)\left(\frac{x}{D}\right)^2 + 2\frac{x}{D} - 1 = 0$$

is the equation for x. Its solution is

$$\frac{x}{D} = \frac{\sqrt{\left(\frac{E}{M}-1\right)}}{\frac{E}{M}-1}.$$

We have $E/M = 81$, and

$$\frac{x}{D} = \frac{8}{80} = \frac{1}{10},$$

so that $x = 3.8 \times 10^4$ km.

CHAPTER 4

(4.1) Let a, b be two arbitrary galaxies which are at *some*
time t^* at equal distances from the origin O. Then

$$r_a(t^*) = r_b(t^*), \text{ i.e. } \bar{R}(t^*)r_a(t_1) = \bar{R}(t^*)r_b(t_1).$$

It follows that $r_a(t_1) = r_b(t_1)$, so that

$$r_a(t) = r_b(t)$$

at *all* times. Since two galaxies are at the same distance from O at one time only if they are at the same
distance from O at *all* times, there can be no overtaking.

(4.2) For constant q, one integrates $-\ddot{R}/\dot{R} = q\dot{R}/R$ to yield, if
A is independent of time,

$$-\ell n\dot{R} = \ell nR^q - \ell nA.$$

Hence

$$\dot{R} = AR^{-q}, \quad H = A/R^{q+1} \qquad (q \neq -1).$$

Integrating again *for* $q = -1$ between t_0 and t where $R(t_0) = 1$, and noting $H = A$,

$$\ell nR(t) = H(t-t_0).$$

Integrating for $q \neq -1$ from $t = 0$ where $R(0) = 0$,

$$R^{q+1} = (q+1)At, \quad H = \frac{1}{(q+1)t}.$$

(4.3)   We have

$$\dot{R}^2 = \frac{C}{R} + \frac{\lambda}{3}R^2 - kc^2,$$

$$\ddot{R} = -\frac{C}{2R^2} + \frac{\lambda}{3}R,$$

$$\dddot{R} = \left(\frac{C}{R^3} + \frac{\lambda}{3}\right)\dot{R}.$$

Multiply these equations by R^{-2}, R^{-1}, $\dot{R}^{-1}$ respectively to find

$$H^2 = \frac{C}{R^3} + \frac{\lambda}{3} - \frac{kc^2}{R^2}, \tag{4.3.1}$$

$$-qH^2 = -\frac{C}{2R^3} + \frac{\lambda}{3}, \tag{4.3.2}$$

$$QH^2 = \frac{C}{R^3} + \frac{\lambda}{3}, \tag{4.3.3}$$

as required.

(4.4)   Equating (4.3.1) and (4.3.2) and also (4.3.1) and (4.3.3) yields

$$\frac{kc^2}{R^2} = \frac{3C}{2R^3} - (q+1)H^2$$

and

$$\frac{kc^2}{R^2} = (Q-1)H^2.$$

Also adding to eqn (4.3.2) multiplied by 2 first (4.3.1),

and then (4.3.3), yields

$$\lambda = (1-2q)H^2 + \frac{kc^2}{R^2}$$

$$\lambda = (Q-2q)H^2.$$

Lastly subtract (4.3.2) from (4.3.3):

$$\frac{3C}{2R^3} = (Q-q)H^2.$$

One sees that $k > 0$ for $Q > 1$, $k < 0$ for $Q < 1$. For $\lambda = 0$ one has that $Q = 2q$. Hence for such models $k > 0$ for $q > \frac{1}{2}$, and $k < 0$ for $q < \frac{1}{2}$.

CHAPTER 5

(5.1) $\sum_{1}^{n} r$ is known to be $\frac{n}{2}(n+1)$, and for large n this yields $n^2/2$. $\sum_{1}^{n} r^2$ is $\frac{n}{6}(n+1)(2n+1)$ and for large n this yields $n^3/3$. The main contributions to these sums comes from the large values of n when the interval between n and $n+1$ is much smaller than n, the term to be added. Under these conditions the sum due to the large n-values is well approximated by an integral. This approximation is poor for the small n-values, but their contribution is negligible anyway. One therefore finds

$$\int_{1}^{n} r \, dr = \frac{n^2-1}{2} \sim \frac{n^2}{2} \text{ for large } n.$$

$$\int_{1}^{n} r^2 \, dr = \frac{n^3-1}{3} \sim \frac{n^3}{3} \text{ for large } n.$$

(5.2) $\sum_{1}^{n} r^{\alpha} \sim \int_{1}^{n} r^{\alpha} \, dr \sim \frac{1}{\alpha+1} n^{\alpha+1}$ for large n and for $\alpha > 0$.

(5.3) $\quad y \equiv \sum_{r=1}^{n} \ln r = \int_{1}^{n} \ln r \ dr = (r \ln r - r)_{1}^{n} \sim n \ln n - n$ for

large n. It follows that for large n

$$e^{y} = e^{\sum_{1}^{n} \ln r} = 1.2.3 \ \ldots \ n = n! = e^{n \ln n} \ e^{-n} = (n/e)^{n}.$$

(5.4)   We chose the time t_0 where the scale factor is unity. Start with a spherical shell of particles whose total mass is $dm(X)$, as given in eqn (5.1). These particles are affected only by the mass $M(X)$ within the radius X. This matter gives rise to a gravitational potential

$$\varphi(X) = -\frac{GM(X)}{X} = -\frac{4\pi}{3}X^3 \rho_0 \frac{G}{X}.$$

Hence the contribution of the spherical shell to the total potential energy is

$$dU(X) \equiv \varphi(X) \ dm(X) = -\frac{4\pi}{3}X^2 \rho_0 G \cdot 4\pi X_0^2 \ dX_0$$

$$= -\frac{16\pi^2}{3}G\rho_0^2 X_0^4 \ dX_0.$$

(5.4.1)

The reason for the first expression is that each element of mass dm can be imagined to be moved from infinity to the surface of radius X. At infinity its potential energy is zero. At the surface it is $\varphi(X) \ dm$. The change in potential energy is $\varphi(X) \ dm - 0 = \varphi(X) \ dm$. Applying this reasoning to all mass elements of the spherical shell of mass given by eqn (5.1), one finds $dU(X) = \varphi(X) \ dm(X)$. The expressions (5.4.1) can now be integrated. One then finds the contribution to the total potential energy from all spherical shells at time t_0

$$V_G(t_0) = -\frac{16\pi^2}{15}G\rho_0^2 a_0^5.$$

(5.4.2)

At a general time t the potential energy $V_G(t) = -B/\bar{R}$

introduced in Section 4.4 behaves as $1/\bar{R}(t)$, so that

$$V_G(t) = V_G(t_0)/\bar{R}(t).$$

Hence

$$B = -V_G(t_0),$$

and this establishes the required result.

CHAPTER 6

(6.1) Use eqn (4.25) to find the density parameter as follows:

$$\sigma = \frac{C}{2H^2R^3} = \frac{C}{2}\cdot\frac{c^2u^6}{c^6k^3(1-u^2)}\cdot\frac{\alpha^3}{u^6}$$

where eqn (6.17) has been used. Since $\alpha = kc^2/C$, one finds

$$\sigma = \frac{1}{2(1-u^2)}.$$

This is q by (6.20). Thus $\sigma = q$, as expected from (4.24) with $\lambda = 0$.

Lastly, consider eqn (4.27). On the left-hand side we have

$$\frac{kc^2}{R^2H^2} = kc^2\cdot\frac{\alpha^2}{u^4}\cdot\frac{c^2u^6}{k^3c^6(1-u^2)} = \frac{u^2}{1-u^2}$$

On the right-hand side we have

$$3\sigma - q - 1 = 2q - 1 = \frac{u^2}{1-u^2},$$

thus confirming (4.27).

(6.2) Near the big bang r and R are small so that $u \ll 1$. In that case

$$f(u) = \frac{\sqrt{(1-u^2)}}{u^3}\{\sin^{-1}u - u\sqrt{(1-u^2)}\}$$

$$= \frac{\sqrt{(1-u^2)}}{u^3}\{u + \frac{1}{6}u^3 + \frac{3u^5}{40} + \ldots - u(1-u^2)^{\frac{1}{2}}\} \doteqdot \frac{2}{3}.$$

$$H \doteqdot \frac{c^3k^{3/2}}{Cu^3} = \frac{c^2k^{3/2}}{C}\frac{c^{3/2}}{(kc^2)^{3/2}R^{3/2}} = \frac{c^{\frac{1}{2}}}{R^{3/2}}$$

$$t = \frac{f(u)}{H} \doteqdot \frac{2}{3}\frac{R^{3/2}}{c^{\frac{1}{2}}} \quad \text{so that} \quad H \doteqdot \frac{2}{3t}$$

For small u, $q \sim 1/2$ by (6.20). Lastly, substituting in $\frac{9}{4}Ct^2/R^3 = 1$ the expression

$$C = \frac{8\pi}{3}G\rho(t)R^3(t)$$

from (5.9), one finds the required result

$$6\pi G\rho(t)t^2 = 1.$$

Alternatively use the equation $\dot{R}^2 = \frac{C}{R} + \frac{1}{3}\lambda R - kc^2$ which for small R leads to the analysis of $\dot{R} = C/R$ as in the Einstein—de Sitter model. The results stated follows then immediately.

(6.3) Use the formula $6\pi G\rho t^2 = 1$ with

$$t = t_0 = 3 \times 10^{17} \text{ s}$$

to find (in cgs units with $G = 6.67 \times 10^{-8}\text{cm}^3\text{gm}^{-1}\text{s}^{-2}$)

$$\rho_0 = \frac{1}{6\pi}\frac{10^8}{6.7}\frac{1}{(3 \times 10^{17})^2} \sim 0.88 \times 10^{-29}\text{gm cm}^{-3}$$

(6.4) By (6.9) with $H_0^{-1} = 3 \times 10^{17}$ s,

$$\rho_c = \frac{3}{8\pi GH_0^{-2}} = \frac{1}{8\pi \times 6.7 \times 10^{-11} \times 9 \times 10^{34}} = 1.98 \times 10^{-26}\text{kg m}^{-3}$$

$$= 1.98 \times 10^{-29}\text{gm cm}^{-3}.$$

(6.5)   The equation of motion is

$$\ddot{x} = -GM/x^2, \text{ i.e. } 2\dot{x}\ddot{x} = -2GM\dot{x}/x^2.$$

Integration leads to

$$\dot{x} = \frac{2GM}{x} + A,$$

where the constant of integration, A, is zero since $\dot{x} = 0$ at $x = \infty$. It follows that

$$\sqrt{x}\, dx = \sqrt{(2GM)}\, dt.$$

Integration, subject to $x = 0$ at $t = 0$, yields

$$x^3 = \frac{9}{2}GMt^2$$

Clearly as $x \to 0$, $\dot{x} \to \infty$.

    The problem corresponds to a cosmological model with $k = 0 = \lambda$, i.e. the Einstein–de Sitter model, provided one makes the identification

$$x \to \text{the scale factor } R$$

$$2GM \to C = \frac{8\pi}{3}G\rho(t)R(t)^3.$$

(6.6)   (i)   The equation of motion and its integration proceed as follows:

$$m\ddot{x} = -GmM/x^2, \quad 2\dot{x}\ddot{x} = -(2GM/x^2)\dot{x}, \quad \dot{x}^2 = 2GM/x + A.$$

The velocity vanishes at $x = x_1$, i.e.

$$A = -2GM/x_1 = v_0^2 - 2GM/x_0. \tag{6.6.1}$$

The energy equation can be written in various ways. The form

$$\dot{x}^2 = \frac{2GM}{x} + v_0^2 - \frac{2GM}{x_0} \tag{6.6.2}$$

shows that the particle will travel to $x = \infty$ if
$v_0^2 > 2GM/x_0$. Then x_1 of (6.6.1) is negative and
the particle will never reach this coordinate
value.

(ii) If $x/x_1 = \sin^2 u/2$ then

$$\dot{u} = \frac{du}{dx}\dot{x} = \frac{\dot{x}}{x_1 \sin(u/2)\cos(u/2)}.$$

Also a form of the energy equation is

$$\dot{x}^2 = \frac{2GM}{x_1}\left(\frac{x_1}{x}-1\right) = \frac{2GM}{x_1}\left[\frac{\cos(u/2)}{\sin(u/2)}\right]^2. \qquad (6.6.3)$$

Hence

$$\dot{u} = \frac{\sqrt{(2GM)}}{x_1^3}\frac{1}{\sin^2 u/2} = \frac{\alpha}{\sin^2 u/2} \qquad (6.6.4)$$

as required, and the energy is

$$E = \frac{1}{2}m\dot{x}^2 - \frac{GmM}{x} = -\frac{GmM}{x_1}. \qquad (6.6.5)$$

(iii) Take $x_1 > 0$. Integrating (6.6.4) with the constraint that $x = 0$ (i.e. $u = 0$) at $t = 0$,

$$\frac{(2GM)}{x_1^3}t = \int_0^u \sin\frac{u}{2}\,du = \frac{1}{2}\int_0^u(1-\cos u)du = \frac{1}{2}(u-\sin u)$$

$$= \frac{u}{2} - \sin\frac{u}{2}\sqrt{(1-\sin^2\frac{u}{2})}$$

$$= \sin^{-1}\sqrt{\left(\frac{x}{x_1}\right)} - \sqrt{\left(\frac{x}{x_1}\right)}\sqrt{\left(1-\frac{x}{x_1}\right)}.$$

The time τ for a complete cycle is given by
$a(t+\tau) = at + 2\pi$, where $a^2 \equiv 2GM/x_1^3$. Thus
$\tau = 2\pi/a$.

Note that for $x_1 > x > 0$ we have $\dot{x}^2 > 0$.
At $x = x_1$ we have $\dot{x} = 0$. If $x > x_1$ there are no
solutions for $\dot{x}$ by (6.6.5), so that x decreases
again after reaching the value $x = x_1$. It
approaches $\dot{x} = \infty$ at $x = 0$.

(iv) Take $x_1 < 0$.

Now $\dot{x}^2$ is always positive by (6.6.5), and the particle goes to infinity. From

$$a \equiv \sin ix = i \sinh x \equiv ib, \ \sin^{-1}(ib)_2 = ix = i \sinh^{-1} b.$$

Hence if v is real then we can put $u = iv$ and

$$-\left|\frac{x}{x_1}\right| = \frac{x}{x_1} = \sin^2 \frac{iv}{2} = -\sinh^2 \frac{v}{2}.$$

Also

$$\sin^{-1}\left(i\left(\sqrt{\left|\frac{x}{x_1}\right|}\right)\right) = i \sinh^{-1}\sqrt{\left(\left|\frac{x}{x_1}\right|\right)}.$$

Hence the result of (ii) becomes

$$-i\frac{\sqrt{(2GM)}}{|x_1|^3}t = \frac{1}{2}(iv - \sinh v)$$

$$= i \sinh^{-1}\sqrt{\left(\left|\frac{x}{x_1}\right|\right)} - i\sqrt{\left(\left|\frac{x}{x}\right|\right)}\sqrt{\left(1 + \left|\frac{x}{x_1}\right|\right)}$$

from which the desired relation follows.

(v) From (6.6.5) one sees that problem (6.5) is recovered when $x_1 \to \infty$.

The equation

$$\dot{x}^2 - \frac{2GM}{x} = \frac{2}{m} E = -\frac{2GM}{x_1}$$

shows that the analogy is brought out by the replacements

$$2GM \to C, \ kc^2 \to -\frac{2}{m}E = \frac{2GM}{x_1}.$$

The case $x_1 > 0$ corresponds to $k = 1$ when one has an oscillating universe. The case $x_1 < 0$ corresponds to $k = -1$ when one has an expanding universe.

(6.7) If $k = 0$ and $\lambda < 0$,

$$R(t)^3 = \left[\frac{3C}{2|\lambda|}\right][1-\cos(\sqrt{(3|\lambda|)}t)].$$

It follows that

$$H = \frac{3R^2\dot{R}}{3R^3} = \frac{[3C/2|\lambda|]\sqrt{(3\lambda)}\sin(\sqrt{(3|\lambda|)}t)}{3[3C/2|\lambda|][1-\cos(\sqrt{(3|\lambda|)}t)]}$$

$$= \frac{\lambda}{3}\frac{\sin(\sqrt{(3|\lambda|)}t)}{1-\cos(\sqrt{(3|\lambda|)}t)}.$$

For small t this yields the Hubble constant for the Einstein—de Sitter model:

$$\sqrt{\left(\frac{|\lambda|}{3}\right)}\frac{\sqrt{(3|\lambda|)}t}{\frac{1}{2}\,3|\lambda|t^2} = \frac{2}{3t}$$

Also one has after two differentiations of the equation for R^3 and multiplying by $1/6\dot{R}R^2$,

$$1 - \tfrac{1}{2}q = \frac{3C}{4}\frac{\cos[\sqrt{(3|\lambda|)}t]}{R\dot{R}^2}.$$

By eqns (6.30) and (6.36)

$$R\dot{R}^2 = C - \frac{C}{2}(1-\cos(\sqrt{(3|\lambda|)})\,) = \frac{C}{2}[1+\cos(\sqrt{(3|\lambda|)}t)].$$

This yields

$$q = \frac{2-\cos(\sqrt{(3|\lambda|)}t)}{1+\cos(\sqrt{(3|\lambda|)}t)}.$$

For small t, $q = \tfrac{1}{2}$.

(6.8) Replacing the density parameter σ by zero and k by -1, eqns (4.24) and (4.27) to (4.28) are

$$-\dot{H} = (1+q)H^2, \quad H^2q = -\lambda/3, \quad (q+1)H^2 = \frac{c^2}{R^2}.$$

In these equations we shall use, with $\alpha^2 = |\lambda|/3$,

$$H = \alpha/\tan\alpha t, \quad q = \tan^2\alpha t, \quad R = (c/\alpha)\sin\alpha t.$$

Then

$$-\dot{H} = \frac{\alpha^2}{\tan^2\alpha t \,\cos^2\alpha t} = H^2\cdot(1+q),$$

as required.

Lastly,

$$(q+1)H^2 = \frac{1}{\cos^2\alpha t}\,\frac{\alpha^2}{\tan^2\alpha t} = \frac{c^2}{R^2}.$$

also as required.

CHAPTER 7

(7.1) For R large enough, $\dot{R}^2 = (\lambda/3)R^2$, so that

$$\frac{\dot{R}}{R} = \sqrt{\frac{\lambda}{3}}, \quad R(t) = R(t_1)\exp\left[\sqrt{\frac{\lambda}{3}}\,(t-t_1)\right]$$

The Hubble constant is

$$H = \frac{\dot{R}}{R} = \sqrt{\frac{\lambda}{3}}.$$

Also $q = -\dfrac{\ddot{R}R}{\dot{R}^2} = -1.$

(7.2) For small enough times the C/R-term dominates, and

$$\dot{R}^2 = \frac{C}{R}, \quad R^3 = \frac{9}{4}Ct^2$$

if $R = 0$ at $t = 0$. It follows that

$$H = \frac{2}{3t}, \quad q = \tfrac{1}{2}.$$

(7.3) By eqn (7.4) with $\alpha = 1$ and $x = 1 + \vartheta$, $|\vartheta| \ll 1$,

$$\ddot{\vartheta} = \beta\left(-\frac{1}{(1+\vartheta)^2}+1+\vartheta\right) = \beta\vartheta \equiv \mu^2\vartheta.$$

This equation has the general solution

$$\vartheta = Ae^{\mu t} + Be^{-\mu t}$$

Similarly eqn (7.3) yields

$$\dot{\vartheta}^2 = \beta\left(\frac{2}{1+\vartheta} + (1+\vartheta)^2 - 3\right) = \beta(2(1-\vartheta)+1+2\vartheta+\vartheta^2-3)$$

so that

$$\dot{\vartheta}^2 \doteq \beta\vartheta^2 \equiv \mu^2\vartheta^2, \quad \dot{\vartheta} = \pm\mu\vartheta.$$

Substituting our general solution into this, we find

$$Ae^{\mu t} - Be^{-\mu t} = \pm(Ae^{\mu t} + Be^{-\mu t})$$

Hence $A = 0$ or $B = 0$, so that one has either an exponential rise or an exponential decay of ϑ, and hence of $(R-R^*)$, with time. The possibility of an exponential rise means that an initial small departure of R from the static value R^* in any direction can lead to an increasingly large departure in that direction, and in this sense the static model is unstable.

CHAPTER 9

(9.1) If $\mu = 0$ in eqn (9.1) one has

$$\ddot{R} = -\frac{4\pi}{3}G\rho R + \frac{1}{3}\lambda R.$$

Multiply by $2\dot{R}$ and integrate to find

$$\dot{R}^2 - (\dot{R}^2)_{t_2} = -\frac{8\pi G}{3}\int_{t_2}^{t} \rho R \; dR + \frac{\lambda}{3}R^2 - \frac{\lambda}{3}(R^2)_{t_2},$$

as required.

(9.2) If one replaces G by G_N in eqn (9.80) and equates to the relativistic result (9.12) one finds

$$\frac{8\pi}{3}G\rho R^2 + \frac{1}{3}\lambda R^2 + \{\dot{R}^2 - \frac{1}{3}\lambda R^2 - \frac{8\pi G}{3}\rho R^2\}_{t_2}$$

$$= -\frac{8\pi}{3}G_N \int_{t_2}^{t} \rho R\ dR + \frac{1}{3}\lambda R^2 + \{\dot{R}^2 - \frac{1}{3}\lambda R^2\}_{t_2}.$$

Multiplying by $3/8\pi G$,

$$\rho R^2 - (\rho R^2)_{t_2} = -\frac{G_N}{G} \int_{t_2}^{t} \rho R\ dR,$$

and the result follows. Thus the energy equations for the two theories do not agree, but the difference can be removed by introducing an 'effective' gravitational constant.

(9.3) For zero pressure models eqn (9.8) holds, i.e. there exists a constant a such that $\rho = aR^{-3}$. Hence eqn (9.31) yields

$$\frac{G_N}{G} = \frac{\dfrac{a}{R_2} - \dfrac{a}{R}}{a \displaystyle\int_{R_2}^{R} R^{-2}\ dR} = 1.$$

(9.4) (i) The energy eqn (9.5) is, with $p = gc^2\rho$,

$$R^3\ d\rho + \rho\ dR^3 + g\rho\ dR^3 = 0,$$

and this may be written as

$$\frac{d\rho}{\rho} = -(1+g)\frac{dR^3}{R^3}, \text{ i.e. } d[\ln \rho] = -d[\ln R^{3+3g}].$$

This shows that $\rho R^{3+3g} = A$ (say) where A is independent of time.

(ii) Substituting this result in (9.31)

$$\frac{G_N}{G} = \frac{\dfrac{A}{R_2^{1+3g}} - \dfrac{A}{R^{1+3g}}}{A\displaystyle\int_{R_2}^{R} R^{-2-3g}\,\mathrm{d}R} = 1 + 3g.$$

(9.5) (i) We have the total differential of G, using the second law in the form (9.20),

$$\mathrm{d}G = \mathrm{d}U + p\mathrm{d}V + V\mathrm{d}p - T\mathrm{d}S - S\mathrm{d}T = T\mathrm{d}S + V\mathrm{d}p - T\mathrm{d}S - S\mathrm{d}T$$

so that

$$\mathrm{d}G = V\mathrm{d}p - S\mathrm{d}T.$$

It follows that

$$\frac{\partial^2 G}{\partial p\,\partial T} = \frac{\partial}{\partial T}\left[\left(\frac{\partial G}{\partial p}\right)_T\right] = \frac{\partial}{\partial V}\left[\left(\frac{\partial G}{\partial T}\right)_V\right].$$

Hence

$$\left(\frac{\partial p}{\partial T}\right)_V = \left(\frac{\partial S}{\partial V}\right)_T.$$

(ii) From the second law for quasistatic processes

$$\mathrm{d}U = T\mathrm{d}S - p\mathrm{d}V.$$

Hence

$$\left(\frac{\partial U}{\partial V}\right)_T \mathrm{d}V + \left(\frac{\partial U}{\partial T}\right)_V \mathrm{d}T = T\left(\frac{\partial S}{\partial V}\right)_T \mathrm{d}V + T\left(\frac{\partial S}{\partial T}\right)_V \mathrm{d}T - p\mathrm{d}V.$$

The coefficients of $\mathrm{d}V$ must be the same on both sides of the equation, so that using the result of (i)

$$\left(\frac{\partial U}{\partial V}\right)_T + p = T\left(\frac{\partial S}{\partial V}\right)_T = T\left(\frac{\partial p}{\partial T}\right)_V.$$

(9.6)   We are to solve

$$U = T\left(\frac{\partial U}{\partial T}\right)_V - \frac{V}{g}\left(\frac{\partial U}{\partial V}\right)_T,$$

in the form

$$U = V^{-g}f(z,T), \quad z \equiv TV^g$$

where f is some function of z and T.   Then

$$\left(\frac{\partial U}{\partial T}\right)_V = V^{-g}\left(\frac{\partial f}{\partial z}\right)_T\left(\frac{\partial z}{\partial T}\right)_V + V^{-g}\left(\frac{\partial f}{\partial T}\right)_z\left(\frac{\partial T}{\partial T}\right)_V = \left(\frac{\partial f}{\partial z}\right)_T + V^{-g}\left(\frac{\partial f}{\partial T}\right)_z$$

$$\left(\frac{\partial U}{\partial V}\right)_T = -\frac{g}{V}U + V^{-g}\left(\frac{\partial f}{\partial z}\right)_T\left(\frac{\partial z}{\partial V}\right)_T + V^{-g}\left(\frac{\partial f}{\partial T}\right)_z\left(\frac{\partial T}{\partial V}\right)_T$$

$$= -\frac{g}{V}U + \frac{gT}{V}\left(\frac{\partial f}{\partial z}\right)_T.$$

The equation for U is therefore

$$U = T\left(\frac{\partial f}{\partial z}\right)_T + TV^{-g}\left(\frac{\partial f}{\partial T}\right)_z + U - T\left(\frac{\partial f}{\partial z}\right)_T.$$

This requires that the function f be independent of T if z is kept constant.   It follows that $UV^g = f(z)$.

One may show similarly that if $U = V^{-g}f(z,V)$ then the function f must be independent of V.

(9.7)   For any simple fluid

$$dU = \left(\frac{\partial U}{\partial V}\right)_T dV + \left(\frac{\partial U}{\partial T}\right)_V dt$$

$$= \left(\frac{\partial U}{\partial V}\right)_T\left[\left(\frac{\partial V}{\partial p}\right)_T dp + \left(\frac{\partial V}{\partial T}\right)_p dT\right] + \left(\frac{\partial U}{\partial T}\right)_V dT$$

$$= \left(\frac{\partial U}{\partial p}\right)_T dp + \left[\left(\frac{\partial U}{\partial V}\right)_T\left(\frac{\partial V}{\partial T}\right)_p + \left(\frac{\partial U}{\partial T}\right)_V\right]dT.$$

Hence

$$\left(\frac{\partial U}{\partial T}\right)_p - \left(\frac{\partial U}{\partial T}\right)_V = \left(\frac{\partial U}{\partial V}\right)_T\left(\frac{\partial V}{\partial T}\right)_p.$$

For an ideal classical gas the right-hand side vanishes by

eqn (9.23).

Now for any simple fluid one has by eqn (9.25)

$$C_p = \left(\frac{\partial U}{\partial T}\right)_p + p\left(\frac{\partial V}{\partial T}\right)_p$$

$$= \left(\frac{\partial U}{\partial T}\right)_v + \left[\left(\frac{\partial U}{\partial V}\right)_T + p\right]\left(\frac{\partial V}{\partial T}\right)_p.$$

But

$$C_V = \left(\frac{\partial U}{\partial T}\right)_V,$$

and this proves the result.

(9.8)   Use $pV = gU$, $TS = jU$ in eqn (9.20):

$$dS = \frac{1}{T}dU + \frac{p}{T}dV = \frac{S}{jU}dU + \frac{gS}{jV}dV = \frac{S}{j}d(\ln\ UV^g).$$

Hence

$$\ln\ S^j = \ln\ UV^g + \text{n}\ j\lambda^{j-1}$$

where the last term is a constant of integration.   Hence

$$UV^g = S^j/j\lambda^{j-1}.$$

Using eqn (9.19)

$$f(z) = S^j/j\lambda^{j-1}.$$

Hence from eqn (9.21)

$$df(z) = \frac{jS^{j-1}}{j\lambda^{j-1}}dS = zdS.$$

It follows that

$$S = \lambda z^{1/(j-1)},$$

so that

$$f(z) = \frac{\lambda}{j} z^{j/(j-1)}$$

For black-body radiation eqn (9.32) shows that $j = \frac{4}{3}$, so that $f(z) \propto z^4$ in agreement with eqn (9.31).

(9.9)   The energy of the system consists of rest energy plus thermal energy and is by eqn (9.28)

$$U = N_m m_0 c^2 + C_V T_m = N_m m_0 c^2 + \frac{p_m V}{\gamma - 1}$$

i.e. $$\rho_m = n_m m_0 + C_V T_m / V c^2 = n_m m_0 + \frac{p_m}{c^2 (\gamma - 1)}$$

Substituting in eqn (9.42) yields eqn (9.43) with $a = 0$. Hence, since $n_m R^3 \propto N_m$ is time independent,

$$\frac{n_m R^3 k}{c^2 (\gamma - 1)} \frac{\mathrm{d}T_m}{\mathrm{d}R} + \frac{3R^2}{c^2} n_m k T_m = 0,$$

so that

$$\frac{\mathrm{d}T_m / \mathrm{d}R}{T_m} = -\frac{3(\gamma - 1)}{R}$$

and

$$T_m R^{3(\gamma - 1)} \text{ is time-independent.}$$

Put $c^{-2} C_V T_m = C V R^{-3\gamma}$; then C is a constant and

$$\rho_m = \frac{m_0 (n_m R^3)}{R^3} + \frac{C}{R^{3\gamma}} \left[\equiv \frac{C_m}{R^3} + \frac{C}{R^5} \text{ if } \gamma = 5/3 \right].$$

This generalizes eqn (9.44) for dust by adding a term (9.46) due to pressure.

(9.10) In this case

$$pV = gU = g(\rho V - N m_0) c^2, \text{ i.e. } p = g(\rho - n m_0) c^2.$$

Eqn (9.42) is

$$R^3 \frac{d\rho}{dR} + 3R^2\rho + \frac{3R^2}{c^2}g(\rho-nm_0)c^2 = 0.$$

Consider the variable $y \equiv \rho R^3 - A$, where A is a constant to be chosen). The above equation for ρ becomes one for g:

$$\frac{dy}{dR} + 3g\left[\frac{y+A}{R} - \frac{m_0 nR^3}{R}\right] = 0.$$

If one choses $A = m_0 nR^3$, which is a constant, the equation for y is

$$\frac{dy}{y} = -3g\frac{dR}{R} \quad \text{i.e.} \quad yR^{3g} \text{ is a constant.}$$

Thus

$$\rho = \frac{m_0(nR^3)}{R^3} + \frac{C}{R^{3(g+1)}}$$

where C is a constant of integration. It follows that

$$p = g(\rho-nm_0)c^2 = \frac{gc^2C}{R^{3(g+1)}},$$

and

$$(\rho-nm_0)R^{3(g+1)} = C.$$

The left-hand side is a statement about the internal energy U of the system (in excess of the rest mass-energy), namely that the following quantity is a constant:

$$\frac{U}{V} \cdot V^{g+1} = UV^g.$$

By (9.22) this states that the entropy of the ideal quantum gas is a constant.

(9.12) $\rho_{mo} = 10^{-30} \text{ gm cm}^{-3} = 10^{-27} \text{ kg m}^{-3}$

$$\rho_{ro} = \frac{a}{c^2}T_{r0}^4 = \frac{7.564 \times 10^{-16} \times (2.70)^4}{(2.998)^2 \, 10^{16}} = 0.447 \, 10^{-30} \text{ kg m}^{-3}$$

Therefore $\qquad\qquad\qquad \rho_{mo}/\rho_{ro} = 2.24 \times 10^3.$

(9.13)   (i)   By (9.33)

$$n_{ro} = 0.3702 \, aT^3/k = \frac{0.3702 \times 7.5641 \times 10^{-16} \times (2.70)^4}{1.3806 \times 10^{-23}}$$

$$= 1.08 \times 10^9 \, m^{-3}.$$

The number of nucleons, based on $\rho_{mo} \sim 10^{-27} kg \, m^{-3}$ is

$$n_{mo} = \frac{10^{-27}}{1.67 \times 10^{-27}} = 0.599 \, m^{-3}.$$

Hence

$$\frac{n_{ro}}{n_{mo}} = 1.803 \times 10^9.$$

(ii)   From eqn (9.31) $U_r \propto T_r^4$, so that $\dfrac{\delta U_r}{U_r} = 4\dfrac{\delta T_r}{T_r}.$

The radiation energy per nucleon is

$$\frac{\rho_{ro} c^2}{n_{mo}} = \frac{0.4473 \times 10^{-30} \times 2.9979^2 \times 10^{16}}{0.599 \times 1.6021 \times 10^{-19}} = 4.189 \times 10^5 \, ev.$$

The energy to ionize a hydrogen atom is $\delta U_r = 13.6 \, ev.$ Hence

$$\frac{\delta T_r}{T_r} = \frac{1}{4} \, \frac{13.6}{4.189 \; 10^5} = 8.116 \; 10^{-6}.$$

(9.14)   Subtract eqn (9.22) from q times (9.20) to find

$$qK - qL - q - L = 2q\sigma - q - \sigma - 3g\sigma.$$

so that

$$qK - (q+1)L = (2q-3g-1)\sigma.$$

(9.15)   A combination of α times eqn (9.22) less β times eqn (9.24) yields

$$(\alpha-\beta)q + \alpha L - \beta K = \alpha\sigma - \beta + (g\alpha-\beta-\beta g)3\sigma.$$

Take $\alpha = 3(1+g)$, $\beta = 1+3g$ when the right-hand side becomes

$$3\sigma + 3g\sigma - 1 - 3g + 3\sigma(-1-g) = -(1+3g).$$

The left-hand side is

$$2q + 3(1+g)L - (1+3g)K.$$

The result follows.

(9.16) Since by (9.14) and (9.18)

$$C = 8\pi G\rho R^3/3$$

$$\sigma = 4\pi G\rho/3H^2$$

the required result follows at once.

CHAPTER 10

(10.1) If $s = 1$ then $E(\nu_1,\nu_2) = \displaystyle\int_{\nu_1}^{\nu_2} \frac{K}{x}dx = K\{\log_e(\nu_2)-\log_e(\nu_1)\}$

$$= K \log_e(\nu_2/\nu_1).$$

From eqn (10.11), therefore,

$$f_\nu = \frac{\log_e(\nu_{2E}/\nu_{1E})}{\log_e(\nu_{20}/\nu_{10})} = 1$$

because from (10.9) $\nu_{2E}/\nu_{1E} = \nu_{20}/\nu_{10}$. (10.13) also implies $f_\nu = 1$ for $s = 1$.

(10.2) (a) From (A.23) and (10.29) the Milne model has

$$(1+z)N_E = N_0\left\{\frac{ct_H}{ct_H-2r}\right\}^2.$$

Hence (10.31) is

$$N_{total} = 4\pi N_0 \int_0^{0.5ct_H} r^2 \left\{ \frac{ct_H}{ct_H - 2r} \right\}^2 dr$$

or if $x \equiv r/ct_H$ and $u \equiv 0.5 - x$,

$$N_{total} = 4\pi N_0 (ct_H)^3 \int_0^{0.5} x^2 \left(\frac{1}{1-2x} \right)^2 dx$$

$$= \pi N_0 (ct_H)^3 \int_0^{0.5} \frac{(0.5-u)^2 du}{u^2}.$$

Near $u = 0$ the integrand behaves like $0.25/u^2$ and the integral diverges.

(b) From (2.8), (2.9), and (10.29) the steady-state model has

$$(1+z)N_E = (1+z)N_0 = \left(\frac{1+r/ct_H}{1-r/ct_H} \right)^{\frac{1}{2}} N_0.$$

Hence (10.31) is

$$N_{total} = 4\pi N_0 \int_0^{ct_H} r^2 \left(\frac{1+r/ct_H}{1-r/ct_H} \right)^{\frac{1}{2}} dr$$

$$= 4\pi N_0 (ct_H)^3 \int_0^1 x^2 \left(\frac{1+x}{1-x} \right)^{\frac{1}{2}} dx$$

$$= 4\pi N_0 (ct_H)^3 \int_0^1 (1-w)^2 \left(\frac{2-w}{w} \right)^{\frac{1}{2}} dw$$

$$< 4\pi N_0 (ct_H)^3 \int_0^1 \left(\frac{2}{w} \right)^{\frac{1}{2}} dw$$

$$= 8\sqrt{2}\ \pi N_0 (ct_H)^3.$$

(10.3) We can take $t_0 = 0$ since the steady-state model has no natural origin of time. Then the equations of motion are:

$$\frac{\mathrm{d}r}{\mathrm{d}t} = c + Hr \text{ for } 0 \leqslant t \leqslant t_{\mathrm{rp}}, \qquad (10.3.1)$$

$$\frac{\mathrm{d}r}{\mathrm{d}t} = -c + Hr \text{ for } t_{\mathrm{rp}} \leqslant t \leqslant t_1 \qquad (10.3.2)$$

and the boundary conditions are

$$r = 0 \text{ at } t = 0 \text{ and at } t = t_1, \ r = r_G \text{ at } t = t_{\mathrm{rp}}.$$
$$(10.3.3)$$

Consider (10.3.1) and define a variable

$$u \equiv r + \frac{c}{H} = r + ct_H \qquad (10.3.4)$$

Then from (10.3.1)

$$\frac{\mathrm{d}u}{\mathrm{d}t} = \frac{\mathrm{d}r}{\mathrm{d}t} = Hr + c = Hu$$

which has the general solution

$$u = Ae^{Ht} \text{ and so } r = Ae^{Ht} - ct_H \text{ for } 0 \leqslant t \leqslant t_{\mathrm{rp}}.$$
$$(10.3.5)$$

The boundary conditions $r = 0$ at $t = 0$ and $r = r_G$ at $t = t_{\mathrm{rp}}$ together with (10.3.5) give

$$A = ct_H \text{ and } r_G = ct_H(\exp(Ht_{\mathrm{rp}})-1). \qquad (10.3.6)$$

In the same way, defining a variable $w \equiv r - ct_H$ reduces (10.3.2) to $\mathrm{d}w/\mathrm{d}t = Hw$ and gives the solution

$$r = Be^{Ht} + ct_H \text{ for } t_{\mathrm{rp}} \leqslant t \leqslant t_1. \qquad (10.3.7)$$

The boundary conditions $r = r_G$ at $t = t_{\mathrm{rp}}$ and $r = 0$ at $t = t_1$ together with (10.3.6) give

$$B = -ct_H \exp(-Ht_1) \text{ and } r_G = ct_H\{1-\exp(Ht_{\mathrm{rp}})\exp(-Ht_1)\}.$$
$$(10.3.8)$$

Now introduce the variable $x \equiv r_G/ct_H$. (10.3.6) gives $\exp(Ht_{\mathrm{rp}}) = 1 + x$ and substituting this in (10.3.8) gives $x = 1 - (1+x) \exp(-Ht_1)$, $\exp(Ht_1) = (1+x)/(1-x)$, and finally $t_1 = t_H \log_e\{(1+x)/(1-x)\}$. Since we have chosen $t_0 = 0$, (10.56) is in the present case

$$d_r = \frac{ct_1}{2} = \frac{ct_H}{2} \log_e\{(1+x)/(1-x)\},$$

as required.

CHAPTER 11

(11.1) Model SC has $r = ct_H z$ and so $d_L = ct_H z (1+z)^2$.

Model MC has $r = ct_H z/(1+z)$ and so $d_L = ct_H z (1+z)$.

Model EC has $r = 2ct_H z/(2+3z)$ and so
$$d_L = ct_H z (1+z)^2/(1+1.5z).$$

Model SSR has $r = ct_H z (2+z)/\{(1+z)^2+1\}$ and so
$d_L = ct_H z (2+z)(1+z)^2/\{(1+z)^2+1\}.$

Model ESR has $r = \dfrac{2ct_H z (2+z)/[(1+z)^2+1]}{2+3z(2+z)/[(1+z)^2+1]}$

$$= \frac{ct_H z (2+z)}{2+2z+z^2+1.5z(2+z)}$$

$$= \frac{ct_H z (1+0.5z)}{1+2.5z+1.25z^2}$$

and $d_L^2 = (1+z)^2 r$.

(11.2) For model MC, eqns (11.36) to (11.39) imply

$$N' = \int_0^{x_L} \frac{x^2(1+z)}{(1-x)^3} \, dx$$

and also

$$1 + z = 1 + \frac{v}{c} = \frac{1}{1-x},$$

so that

$$N' = \int_0^{x_L} \frac{x^2 dx}{(1-x)^4} = \int_{1-x_L}^1 \frac{(1-u)^2 du}{u^4} \qquad (u \equiv 1-x)$$

$$= \int_{1-x_L}^1 (u^{-4} - 2u^{-3} + u^{-2}) du$$

$$= \left| -\frac{1}{3u^3} + \frac{1}{u^2} - \frac{1}{u} \right|_{1-x_L}^1$$

$$= -\frac{1}{3} + \frac{1}{3(1-x_L)^3} - \frac{1}{(1-x_L)^2} + \frac{1}{1-x_L}$$

$$= \frac{1}{3(1-x_L)^3} \{-(1-x_L)^3 + 1 - 3(1-x_L) + 3(1-x_L)^2\}$$

$$= \frac{x_L^3}{3(1-x_L)^3},$$

as required. Also from eqns (11.36) to (11.40) and $z = v/c$,

$$S' = (1+z_L)^{-4}/x_L^2 = (1-x_L)^4/x_L^2.$$

For model MSR, eqns (11.36) to (11.39) imply

$$N' = \int_0^{x_L} \frac{x^2(1+z)}{(1-2x)^{3/2}} dx,$$

and also

$$1 + z = \left\{ \frac{1+v/c}{1-v/c} \right\}^{\frac{1}{2}} = \frac{1}{(1-2x)^{\frac{1}{2}}}, \qquad (11.2.1)$$

so that

$$N' = \int_0^{x_L} \frac{x^2 dx}{(1-2x)^2} = \int_{1-2x_L}^1 \frac{(1-w)^2 dw}{8w^2}$$

$$= \frac{1}{8} \int_{1-2x_L}^{1} (w^{-2} - 2w^{-1} + 1)\, dw$$

$$= \frac{1}{8} \left| (-w^{-1} - 2 \ln w + w) \right|_{1-2x_L}^{1}$$

$$= \frac{1}{8} \left\{ \frac{1}{1-2x_L} + 2 \ln(1-2x_L) - (1-2x_L) \right\},$$

as required. Also from eqns (11.40) and (11.2.1)

$$S' = (1-2x_L)^2 / x_L^2.$$

For model SSR

$$1 + z = \left\{ \frac{1+v/c}{1-v/c} \right\}^{\frac{1}{2}} = \left\{ \frac{1+x}{1-x} \right\}^{\frac{1}{2}}$$

so eqns (11.37) and (11.39) imply directly

$$N' = \int_{0}^{x_L} x^2 \left\{ \frac{1+x}{1-x} \right\}^{\frac{1}{2}}\, dx$$

and eqn (11.40) implies

$$S' = \left\{ \frac{1-x_L}{1+x_L} \right\}^2 x_L^{-2} = \left\{ \frac{1-x_L}{x_L + x_L^2} \right\}^2$$

(11.3) Substituting (11.8) and (11.10) into (11.47) gives

$$N(S) = 4\pi c N_0 \int_{t_E}^{0} \left(\frac{c}{H} \right)^2 \{\exp(Ht_E) - 1\}^2\, dt_E.$$

Since for this model $H = H_0 = 1/t_H$, we can introduce the variable $x \equiv t_E/t_H$ to give

$$N' = \int_{x_L}^{0} \{e^x - 1\}^2\, dx \qquad (x_L \equiv t_L/t_H)$$

$$= \int_{x_L}^{0} \{e^{2x} - 2e^x + 1\}\, dx$$

$$= \left| \left(\frac{e^{2x}}{2} - 2e^{x} + x \right) \right|_{x_{L}}^{0}$$

$$= -1.5 - \frac{1}{2w^{2}} + \frac{2}{w} + \ln(w),$$

as required, if $w \equiv e^{-x_{L}} = \exp(-t_{L}/t_{H}) = \exp(-Ht_{L})$.
Substituting eqns (11.9) to (11.11) into eqn (11.49)
also leads to

$$S' = \{\exp(-Ht_{L})\}^{-2} \{\exp(-Ht_{L}) - 1\}^{-2} = w^{-2}(w-1)^{-2}$$

as required.

(11.4)   From eqn (11.15)

$$s = 1 - (t_{E}/t_{0})^{1/3}.$$

Using this and eqn (11.13), eqn (11.48) becomes

$$N(S) = 4\pi c N_{0}(3ct_{0})^{3} \int_{t_{L}}^{t_{0}} \frac{\{1 - (t_{E}/t_{0})^{1/3}\}^{2} \, dt_{E}}{3ct_{0}^{1/3} t_{E}^{2/3}}.$$

Introducing $y \equiv (t_{E}/t_{0})^{1/3}$, so that $dt_{E} = 3t_{0}y^{2} \, dy$,
this becomes

$$N(S) = 4\pi N_{0}(3ct_{0})^{3} \int_{y_{L}}^{1} \{1-y\}^{2} \, dy$$

and since $3ct_{0} = 2ct_{H}$ for Einstein—de Sitter models, we
have

$$N' = 8 \int_{y_{L}}^{1} \{1-y\}^{2} \, dy = \frac{8}{3}(1-y_{L})^{3}.$$

From eqn (11.14) $R(t_{0}) = 2ct_{H}$. Using this, eqn (11.15)
and the equation following (11.15), eqn (11.49) implies

$$S' = \frac{(1-s_L)^4}{s_L^2} = \frac{(t_L/t_0)^{4/3}}{4(1-(t_L/t_0)^{1/3})^2} = \frac{y_L^4}{4(1-y_L)^2},$$

as required.

(11.5) The integrand in eqn (11.39) is to be multiplied by $S = A/(ct_H)^2 x^2 (1+z)^4$ and the upper limit extended to a 'horizon' distance or maximum distance $x_h = 1$(MC) or $2/3$(EC). For MC, using $z = v/c$ and the results of eqns (11.36) and (11.37), we find the total incident energy as

$$E = 4\pi A N_0 ct_H \int_0^1 \frac{dx}{(1+z)^3(1-x)^3}$$

$$= 4\pi A N_0 ct_H \int_0^1 dx = 4\pi A N_0 ct_H$$

and so r(equiv.) $= 1$.

    Similarly for EC

$$E = 4\ AN_0 ct_H \int_0^{2/3} \frac{dx}{(1+z)^3(1-1.5x)^2}$$

and so

$$r(\text{equiv.}) = \int_0^{2/3} \frac{dx}{\left\{\frac{1-0.5x}{1-1.5x}\right\}^3 (1-1.5x)^2} = \int_0^{2/3} \frac{(1-1.5x)\ dx}{(1-0.5x)^3}$$

$$= \left| \frac{2}{u^2} - \frac{6}{u} \right|_{2/3}^1 = 0.5 \quad (u \equiv 1 - 0.5x).$$

(11.6) The same procedure is used here as in Problem 11.5, but with $x_h = 1$(SSR) and 0.5(MSR), and with $1 + z = \left\{\frac{1+v/c}{1-v/c}\right\}^{\frac{1}{2}}$.

For model SSR, using eqns (11.36) and (11.37), we find

$$E = 4\pi A N_0 ct_H \int_0^1 \frac{dx}{(1+z)^3}$$

so

$$r(\text{equiv.}) = \int_0^1 \frac{dx}{(1+z)^3} = \int_0^1 \left(\frac{1-x}{1+x}\right)^{3/2} dx = 0.2876.$$

Similarly, for model MSR we find

$$E = 4\pi AN_0 c t_H \int_0^{0.5} \frac{dx}{(1-2x)^{3/2}(1+z)^3}$$

$$= 4\pi AN_0 c t_H \int_0^{0.5} dx = 2\pi AN_0 c t_H$$

so that $r(\text{equiv.}) = 0.5$.

(11.7)   In model SGR the total energy incident on a sphere of
unit cross-section is found by multiplying the integrand
in eqn (11.47) by the brightness S, and extending the
lower limit of integration to $-\infty$. Since
$S = A/R^2(t_0)(1+z)^2 s^2$ and $R(t_0) = (1+z)R(t_E)$ (from eqns
(10.45) and (10.43)) the result is (with $t_0 = 0$)

$$E = 4\pi c AN_0 \int_{-\infty}^0 \frac{dt_E}{(1+z)^4}$$

and so $r(\text{equiv.}) = \dfrac{E}{4\pi AN_0 c t_H} = \dfrac{1}{t_H} \displaystyle\int_{-\infty}^0 \frac{dt_E}{(1+z)^4}.$

Using eqn (11.11) for $1+z$, and also the fact that
$t_H = 1/H$, we find

$$r(\text{equiv.}) = \int_{-\infty}^0 \exp(4Ht_E)H dt_E = 0.25.$$

In model MGR the same procedure is used, starting with
eqn (11.48) and extending the lower limit of integration
to zero. The result is

$$E = 4\pi c AN_0 \int_0^{t_0} \frac{dt_E}{(1+z)} = \frac{4\pi c AN_0}{t_0} \int_0^{t_0} t_E dt_E$$

(using $1 + z = R(t_0)/R(t_E) = t_0/t_E$ from eqn (11.18)).
Since $t_0 = t_H$ for Milne models this implies

$$r(\text{equiv.}) = \frac{1}{t_0^2} \int_0^{t_0} t_E dt_E = 0.5.$$

APPENDIX A

(A.1)   If $v = 0.6c$ then from eqn (A.2) $\beta = 1.25$, so by eqn
(A.3) a time interval of 1 year on the ship corresponds
to 1.25 years in S. The ship therefore moves a distance
$(0.6 \times 1.25) = 0.75$ light years in S between emitting
successive signals, and the travel time of a signal
emitted at distance x light-years from 0 is x years.
Labelling the emission of the ith signal by E_i and its
observation at 0 by O_i, the following table is easily
checked, and implies the results asked for.

Number of signal, i	1	2	3	4	5	6	7	8
$x_{Ei} \equiv$ x-coordinate of E_i	0.75	1.5	2.25	3	2.25	1.5	0.75	0
$t_{Ei} \equiv$ t-coordinate of E_i	1.25	2.5	3.75	5	6.25	7.5	8.75	10
$t_{Oi} \equiv$ t-coordinate of O_i	2.0	4.0	6.0	8.0	8.5	9	9.5	10

Note that if $v = 0.6c$, $1 + z = \left\{\dfrac{1.6}{0.4}\right\}^{\frac{1}{2}} = 2$, so the time
interval between observations of signals emitted on the
outward journey is 2 years, in agreement with the table.
Similarly, if $v = -0.6c$, $1 + z = 0.5$.

(A.2)   Eqn (10.3) remains valid for a galaxy moving in a general direction, and must be supplemented by $\Delta r = v_r \Delta t_E =$
$v \cos \vartheta\, \Delta t_E$, where $v_r = v \cos \vartheta$ is the radial component
of velocity of the galaxy. It follows at once from
(10.3) that $\Delta t_0 = \Delta t_E[1 + (v \cos \vartheta)/c]$. However, the direction of motion is irrelevant to the time dilation factor
of eqn (A.3), so that eqn (A.4) gives

$$1 + z = \frac{\Delta t_0}{\Delta t_E} \frac{\Delta t_E}{\Delta' t_E} = \beta \left(1 + \frac{v \cos \vartheta}{c}\right)$$

as required. If $\vartheta = 0$ then $\cos \vartheta = 1$ and (A.5) is recovered. If $\vartheta = 180°$ then $\cos \vartheta = -1$ and

$$1 + z = \frac{1 - v/c}{(1 - v^2/c^2)^{\frac{1}{2}}} = \frac{1 - v/c}{\{(1 - v/c)(1 + v/c)\}^{\frac{1}{2}}} = \left\{\frac{1 - v/c}{1 + v/c}\right\}^{\frac{1}{2}}$$

which is eqn (A.6) with the sign of v reversed.

The condition for zero red-shift is

$$1 = \beta^2 \left(1 + \frac{v \cos \vartheta}{c}\right)^2 \quad \text{or if } y \equiv v/c \text{ and } x \equiv \cos \vartheta,$$

$$1 = \frac{1}{(1-y^2)}(1+xy)^2;$$

$$1 - y^2 = 1 + 2xy + x^2 y^2;$$

$$yx^2 + 2x + y = 0.$$

This is a quadratic in x, with solution (since $|x| \leqslant 1$)

$$x = \frac{-1 + \sqrt{(1-y^2)}}{y}$$

and $v = 0.8c$ gives $\cos \vartheta = \frac{-1 + 0.6}{0.8} = -0.5$, so $\vartheta = 120°$.

(A.3)   The maximum red-shift occurs when motion is radially away from the observer ($\vartheta = 0$ in the notation of Problem (A.2)) and is z_{max}, where

$$1 + z_{max} = \left(\frac{1+v/c}{1-v/c}\right)^{\frac{1}{2}}.$$

(i)   Since $z_{max} = 2$, it follows that $1 + v/c = 9(1-v/c)$, $10v/c = 8$,

$$v = 0.8c.$$

(ii)   The maximum blue-shift occurs when motion is towards the observer ($\vartheta = 180°$), and its magnitude is z_{bmax}, where

$$1 - z_{bmax} = \left(\frac{1-v/c}{1+v/c}\right)^{\frac{1}{2}} = \frac{1}{1+z_{max}} = 1/3$$

so that $z_{bmax} = 2/3$.

(iii)   The minimum apparent brightness occurs for $\vartheta = 0$ and is given by eqn (10.20) with eqn (10.29);

$$S_{min} = \frac{A}{(1+z_{max})^4 r^2} = \frac{A}{r^2}\left(\frac{1-v/c}{1+v/c}\right)^2.$$

The maximum apparent brightness occurs for $\vartheta = 180°$ and is obtained by replacing v by $(-v)$, so that

$$S_{max} = \frac{A}{r^2}\left(\frac{1+v/c}{1-v/c}\right)^2.$$

Therefore

$$S_{max}/S_{min} = \left(\frac{1+v/c}{1-v/c}\right)^4 = 9^4 = 6561.$$

APPENDIX B

(B.1)   The table is obtained from formula (B.1) by simple algebra.

(B.2)   The period available for the measurement of the mass difference is the period T since the big bang, $\Delta t \sim T \sim H^{-1}$. The smallest mass difference which can be measured is given by

$$c^2 \Delta m = \hbar H$$

whence $\Delta m = m(-6)$.

(B.3)   Using $[e] = \left[M^{\frac{1}{2}}L^{3/2}T^{-1}\right]$,

$$m = \hbar^{\vartheta} H^{\beta} G^{\gamma} c^{\delta} e^{\varepsilon}$$

$$= M^{\vartheta-\gamma+\varepsilon/2} L^{2\vartheta+3\gamma+\delta+3\varepsilon/2} T^{-\vartheta-\beta-2\gamma-\delta-\varepsilon}.$$

Hence

$$\vartheta = \frac{1}{5}(3-\delta-3\varepsilon), \quad \beta = \frac{1}{5}(1-\varepsilon-2\delta), \quad -\gamma = \frac{1}{5}(2+\delta+\frac{\varepsilon}{2}).$$

Writing $\delta = b/3$, $\varepsilon = 2b'$ in

$$m^5 = \frac{\hbar^3 H}{G^2} \left(\frac{c^5}{\hbar H^2 G}\right)^{\delta + \epsilon/2} \alpha^{5\epsilon/2}$$

yields the required result. The time dependence is governed by

$$m^5 \propto G^{-(1+b+3b')}$$

so that $b' = -(1+b)/3$ for the time-independent masses. Hence

$$m = \left(\frac{\hbar^3 H}{G^2}\right)^{1/5} \left(\frac{c^5}{\hbar H^2 G}\right)^{\frac{b}{15}+\frac{b'}{5}} \alpha^{b'}$$

becomes $m(-1)\alpha^{b'}$.

(B.4)   Let us put

$$[e] = [\hbar^\alpha H^\beta c^\gamma]$$

$$= [(ML^2 T^{-1})^\alpha T^{-\beta} L^\gamma T^{-\gamma}]$$

$$= [M^\alpha L^{2\alpha+\gamma} T^{-\alpha-\beta-\gamma}].$$

For this to be an electric charge,

$$[e] = M^{\frac{1}{2}} L^{3/2} T^{-1}$$

one needs

$$\alpha = \frac{1}{2}, \quad 2\alpha + \gamma = \frac{3}{2}, \quad \alpha + \beta + \gamma = 1.$$

It follows that $\beta = 0$, making the charge time independent. Also $\gamma = \frac{1}{2}$.   Hence

$$e \sim (\hbar c)^{\frac{1}{2}} \text{ or } e^2 \sim \hbar c.$$

The coefficient of proportionality, k, is in this case just the fine-structure constant $\alpha = 1/137$.

(B.5) The quantity to be considered involves e^2, which can always be eliminated from considerations based on the text by writing it as $e^2 = \alpha\hbar c$, i.e. in terms of the fine-structure constant. Hence we have

$$t \equiv \frac{H^{-1}}{e^2/mc^3} = \frac{mc^3}{e^2 H} = \frac{1}{\alpha}\frac{mc^3}{\hbar c H} = \frac{1}{\alpha}\frac{m(-1)}{\hbar H/c^2} = \frac{1}{\alpha}\frac{m(-1)}{m(-6)}.$$

It is an algebraic consequence of (B.1) that

$$N \equiv \frac{m(9)}{m(-1)} = \left[\frac{m(-1)}{m(-6)}\right]^2$$

Hence

$$t = \sqrt{N}/\alpha \sim 67.1 \times 10^{41}.$$

(B.6) We have to replace the reduced electron mass by $\frac{1}{2}m_n$, as explained in the text, and e^2 is replaced by Gm_n^2. We ignore the difference between electron and neutron mass, by using $m(-1)$ for both and the gravitational radius of a stable particle is

$$r_G = \frac{2Gm^2}{mc^2} \sim \frac{2Gm(-1)}{c^2}.$$

From eqns (B.6) and (B.7)

$$r_h = \frac{c}{H} = c\frac{Gm(9)}{c^3}$$

Hence

$$\frac{r_G}{r_h} \sim \frac{m(-1)}{m(9)} \sim \frac{1}{N}.$$

(B.7) The time t_p is given by $m(\rho) \sim m(-1)$. This implies

$$t_p = \frac{1}{H(t_p)} \sim \left[\frac{\hbar G(t_p)}{c^5}\right]^{\frac{1}{2}}$$

Hence

$$m\left(\frac{3}{2}\right)c^2 t_p = \left|\frac{\hbar c}{G(t_p)}c^4\left|\frac{\hbar G(t_p)}{c^5}\right|^{\frac{1}{2}}\right. = \hbar.$$

APPENDIX C

(C.1)   We have to solve the cubic equation

$$y^3 + 3py + 2t = 0 \qquad\qquad (C.1.1)$$

with

$$p \equiv -(3\sigma)^2, \quad t \equiv \tfrac{27}{2}\sigma^2(1-2\sigma), \quad y = q + 1 - 3\sigma. \quad (C.1.2)$$

The nature of the roots of the reduced eqn (C.1.1) is determined by the values of

$$p^3 + t^2 = \tfrac{3^6}{4}\sigma^4(1-4\sigma). \qquad\qquad (C.1.3)$$

   Irreducible case: $\sigma > \tfrac{1}{4}$. Here (C.1.3) is negative
and there are then known to be three real solutions,
y_1, y_2, y_3 say, which can be given in terms of an angle
φ, defined by

$$\cos \varphi = -t/\sqrt{(-p^3)} = 1 - 1/2\sigma.$$

Then

$$y_1 = 2\sqrt{(-p)} \cos(\varphi/3) = 6\sigma \cos(\varphi/3),$$

$$y_2 = -6\sigma \cos(\varphi/3 + \pi/3), \quad y_3 = -6\sigma \cos(\varphi/3 - \pi/3).$$

The values of q which are implied are $q_j = y_j - 1 + 3\sigma$
($j = 1,2,3$). For q^* in eqn (C.13) one must choose that
value which ensures $q > q^*$, i.e. the least of the three
values q_1, q_2, and q_3. This is achieved by

$$q^* = F(\sigma) = q_3 = 3\sigma - 1 - 6\sigma \cos(\varphi/3 - \pi/3).$$

This is the first of the results (C.14).

   Case $\sigma \leqslant \tfrac{1}{4}$. If $p^3 + t^2 > 0$ there are one real and

two complex roots. Only the real root is needed and is
given by

$$y_1 = [-t + \sqrt{(t^2 + p^3)}]^{1/3} + [-t - \sqrt{(t^2 + p^3)}]^{1/3}.$$

Substituting from (C.1.2) yields the solution in the form

$$q^* = F(\sigma) = q_1 = y_1 - 1 + 3\sigma.$$

This leads to a statement of the form
is used.

   Case $\sigma = \dfrac{1}{4}$. The inequality becomes

$$q > q^* = 2(-t)^{1/3} + 3\sigma - 1 = -2t^{1/3} + 3\sigma - 1 = -\frac{7}{4}.$$

(C.2) *Case* $\sigma \geqslant 1$. The angle φ in the irreducible case is now
small, i.e.

$$\cos \varphi \doteqdot 1 - \frac{\varphi^2}{2} = 1 - \frac{1}{2\sigma},$$

so that

$$\varphi = \pm(1/\sigma)^{1/2} \qquad\qquad\qquad (C.2.1)$$

Also

$$F(\sigma) = 3\sigma\left\{1 - 2\cos\frac{\varphi}{3}\cos\frac{\pi}{3} + 2\sin\frac{\varphi}{3}\sin 60\right\} - 1$$

$$= 3\sigma\left\{1 - 2(1 - \frac{\varphi^2}{9})\frac{1}{2} + 2\frac{\varphi}{3}\frac{\sqrt{3}}{2}\right\} - 1.$$

This becomes for small φ and large σ, using (C.2.1)

$$F(\sigma) = \sqrt{3\sigma}\varphi - 1 = \pm\sqrt{(3\sigma)} - 1 \doteqdot -\sqrt{(3\sigma)}.$$

Since $F(\sigma) < q < -1$, the negative sign has to be chosen.
This yields the required result.

(C.3)   From eqns (C.15) and (C.17) one has for $\sigma > q$, and
writing $b \equiv 2\sigma - 1$ and $x \equiv by$,

$$Ht > \int_0^1 \frac{dy}{\sqrt{(2\sigma y^{-1} - b)}} = \int_0^1 \frac{\sqrt{y}\ dy}{\sqrt{(2\sigma - by)}} = b^{-3/2} \int_0^b \frac{\sqrt{x}\ dx}{\sqrt{(2\sigma - x)}}.$$

It follows that, provided $2\sigma > 1$,

$$Ht > (2\sigma - 1)^{-3/2} \int_0^{2\sigma - 1} \frac{\sqrt{x}\ dx}{\sqrt{(2\sigma - x)}}. \tag{C.3.1}$$

We now prove that

$$I(a^2) \equiv \int_0^{a^2 - 1} \frac{\sqrt{x}\ dx}{\sqrt{(a^2 - x)}} = a^2 \left[\frac{\pi}{2} - \sin^{-1} \frac{1}{a} \right] - \sqrt{(a^2 - 1)} \quad (a \geqq 1)$$

Proof.   Let $t^2 \equiv a^2 - x$, then

$$\frac{\sqrt{x}\ dx}{\sqrt{(a^2 - x)}} = \frac{1}{t} \cdot \sqrt{(a^2 - t^2)} \cdot (-2t\ dt).$$

Hence

$$I(a^2) = 2 \int_1^a \sqrt{(a^2 - t^2)}\ dt = \left[t\sqrt{(a^2 - t^2)} + a^2 \sin^{-1} \frac{t}{a} \right]_1^a \tag{C.3.2}$$

Combination of (C.3.1) and (C.3.2) with $a^2 = 2\sigma$ yields (C.18)
for $q < \sigma$.   The cases $q = \sigma$ and $q > \sigma$ are treated similarly.

(C.4)   By eqns (C.3), (C.24), and (C.27)

$$\sigma_{0max} - q_{0min} \equiv \left(\frac{\lambda}{3H_0^2} \right)_{max} = 28.4 - (-10.1) = 38.5$$

$$\sigma_{0min} - q_{0max} \equiv \left(\frac{\lambda}{3H_0^2} \right)_{0\ min} = 0 - 2 = -2$$

$$-2 < \frac{\lambda}{3H_0^2} < 38.5.$$

Since H_0 is positive, it follows that

$$-6\ H_{0max}^2 < \lambda < 115.5\ H_{0max}^2.$$

Since $H_0^{-1} < 19.4 \times 10^9$ years, $H_0 > 5.155 \times 10^{-11}$ years^{-1} and

$$H_0^2 > 26.57 \times 10^{22} \text{ years}^{-2}$$

$$= \frac{26.57 \times 10^{-22}}{9.986 \times 10^{14}} \text{ s}^{-2} = 2.661 \times 10^{-36} \text{ s}^2 \ .$$

Thus

$$\left(H_0^2\right)_{min} = 2.661 \times 10^{-36} \text{ s}^{-2}, \quad \left(H_0^2\right)_{max} = 10.64 \times 10^{-36} \text{ s}^{-2}$$

From this one sees that

$$-6.38 \times 10^{-35} \text{ s}^{-2} < \lambda < 12.30 \times 10^{-34} \text{ s}^{-2}.$$

(C.5) By eqn (C.4)

$$\frac{kc^2}{R^2} = (3\sigma - q - 1)H^2 \ .$$

It follows that

$$3\sigma_{0max} - q_{0min} - 1 = \left(\frac{kc^2}{R_0^2 H_0^2}\right)_{max} = 85.2 + 10.1 - 1 = 94.3$$

$$3\sigma_{0min} - q_{0max} - 1 = \left(\frac{kc^2}{R_0^2 H_0^2}\right)_{min} = -3.$$

As in Problem (C.4) it follows that

$$-3 \, H_{0max}^2 < \frac{kc^2}{R_0^2} < 94.3 \, H_{0max}^2$$

so that

$$-0.3192 \times 10^{-34} \text{ s}^{-2} < \frac{kc^2}{R_0^2} < 10.03 \times 10^{-34} \text{ s}^{-2}.$$

(C.6) We have by eqn (C.2)

$$\sigma_0 = 4\pi G \rho_0(t)/3H_0^2$$

where $G = 6.67 \times 10^{-11}$ m^3 kg^{-1} s$^{-2} = 6.67 \times 10^{-8}$ cm^3 gm^{-1} s^{-2}. Take H_0^{-1} as 10^{10} years $= 3.15 \times 10^{17}$ s. Then one finds that for dense clusters

$$\rho_0(t) = \frac{3H_0^2\sigma_0}{4\pi G} = \frac{3 \times 28.4}{4\pi(3.15)^2 10^{34} \times 6.67 \times 10^{-3}} = 1.02 \times 10^{-27} \text{gm cm}^{-3}.$$

(C.7) (i) Verify that for $\sigma_0 = 28.4$ one finds $\varphi_0 = 0.188$ and

$$2 \cos \frac{\varphi_0 - \pi}{3} \sim 1.106.$$

It follows that

$$q_0 > F(28.4) = (85.2) \times (-0.106) - 1 = -10.03$$

in agreement with eqn (C.25).

(ii) One finds, assuming $q < 28.4 = \sigma$,

$$j(28.4) = \frac{56.8}{(55.8)^{3/2}}[1.57 - 0.134] - 0.018 = 0.178.$$

It follows that $H_0 t_0 > 0.178$, and any *assumed* lower bound for $H_0 t_0$ represents a further constraint only if it exceeds this value. In eqn (C.26) it is assumed that $H_0 t_0 > 0.52$, which is a genuinely additional constraint to what one can deduce from the assumption $q_0 < \sigma_0 = 28.4$.

(C.8) By (C.14) $\cos \varphi = 0$ when $2\sigma = 1$, so that

$$F(\sigma) = F\left(\frac{1}{2}\right) = \frac{3}{2}\{1 - 2 \cos 30°\} - 1 = -\frac{1}{2}[3\sqrt{3} - 1] = -2.098.$$

Hence $q > -2.1$ by (C.13).

(C.9) Approximate $F(\sigma)$ by working out the appropriate expressions for the various parts of $F(\sigma)$. Thus

$$X \equiv \left(\frac{1}{4\sigma^2}-\frac{1}{\sigma}\right)^{1/2} = \frac{1}{2\sigma}(1-4\sigma)^{1/2} \doteqdot \frac{1}{2\sigma} - 1 - \sigma - \cdots$$

Hence

$$Y \equiv \left(1-\frac{1}{2\sigma}+X\right)^{1/3} + \left(1-\frac{1}{2\sigma}-X\right)^{1/3} = (-\sigma)^{1/3} + \left(2-\frac{1}{\sigma}+\sigma\right)^{1/3}$$

$$= \left(-\frac{1}{\sigma}+2\right)^{1/3} = -\frac{1}{\sigma^{1/3}}\left(1-\frac{2}{3}\sigma\right).$$

Hence

$$F(\sigma) = 3\sigma(1+Y) - 1 = 3\sigma\left[1-\frac{1}{\sigma^{1/3}}+\frac{2}{3}\sigma^{2/3}\right] - 1$$

$$= 3\sigma - 3\sigma^{2/3} + 2\sigma^{5/3} - 1.$$

The terms with powers of σ in excess of $2/3$ can be neglected.

(C.10) $H_0 = 100$ km s^{-1} Mpc$^{-1} = \dfrac{10^2 \times 10^5}{3.086 \times 10^{24}} = 0.324 \times 10^{-17}$ s^{-1}.

1 year $= 3.154 \times 10^7$ s.

$H_0 = 0.324 \times 3.154 \times 10^{-10}$ years$^{-1} = 1.022 \times 10^{-10}$ years^{-1}

$H_0^{-1} = 0.978 \times 10^{10}$ years.

(C.11)   The preceding problem may be used to infer, from H_0 in km s^{-1} Mpc^{-1}, $\lambda = 3H_0^2 L_0$, and $kc^2/R_0^2 = K_0 H_0^2$, that

$$0.5039 \times 10^{-35} L_{0min} < \lambda < 3.149 \times 10^{-35} L_{0max} \text{ s}^{-2} \quad (L_{0min} > 0)$$

$$0.1680 \times 10^{-35} K_{0min} < \frac{kc^2}{R_0^2} < 1.05 \times 10^{-35} K_{0max} \text{ s}^{-2} \quad (K_{0min} > 0)$$

$$3.149 \times 10^{-35} L_{0min} < \lambda < 3.149 \times 10^{-35} L_{0max} \text{ s}^{-2} \quad (L_{0min} < 0)$$

$$1.05 \times 10^{-35} K_{0min} < \frac{kc^2}{R_0^2} < 1.05 \times 10^{-35} K_{0max} \text{ s}^{-2} \quad (K_{0min} < 0)$$

The table can be constructed from these formulae.

(C.12)   By eqn (C.2)

$$\rho(t) = \frac{3}{4\pi G}H^2\sigma.$$

One now substitutes the smallest and the largest value of $H^2\sigma$ into this formula.

NOTES AND REFERENCES

CHAPTER 1

1. P.A.M. Dirac (1973). Long range forces and broken symmetries. *Proc. R. Soc.* **A333**, 403.

2. F. Hoyle and J.V. Narlikar (1971). On the nature of mass. *Nature, Lond.* **233**, 41.

3. G. Gamow (1946). The expanding universe and the origin of the elements. *Phys. Rev.* **70**, 572.

4. A.A. Penzias and R.W. Wilson (1965). A measurement of excess antenna temperature at 4080 Mc/s. *Astrophys. J.* **142**, 419.

5. C.B. Collins and S.W. Hawking (1973). The rotation and distortion of the universe. *Mon. Not. R. astr. Soc.* **162**, 307.

6. S.A. Colgate (1974). The formation of deuterium and the light elements by spallation in supernova shocks. *Astrophys. J.* **187**, 321.

7. C. Brans and R.H. Dicke (1961). Mach's principle and a relativistic theory of gravitation. *Phys. Rev.* **124**, 925.

8. N.T. Bishop (1976). Cosmology and a general scalar-tensor theory of gravitation. *Mon. Not. R. astr. Soc.* **176**, 241.

9. P.T. Landsberg and N.T. Bishop (1975). A principle of impotence allowing for Newtonian cosmologies with a time-dependent gravitational constant. *Mon. Not. R. astr. Soc.* **171**, 279.

10. For a recent review *see* P.S. Wesson (1973). The implications for geophysics of modern cosmologies in which G is variable. *Q. Jl R. astr. Soc.* **14**, 9.

CHAPTER 2

1. J.R. Gott, J.E. Gunn, D.N. Schramm, and B.M. Tinsley (1976). Will the universe expand forever? *Scient. Amer.* March 1976.

2. H. Bondi and T. Gold (1948). The steady state theory of the expanding universe. *Mon. Not. R. astr. Soc.* **108**, 252.

3. F. Hoyle (1948). A new model for the expanding universe. *Mon. Not. R. astr. Soc.* **108**, 372.

4. E.A. Milne (1935). *Relativity, gravitation, and world structure,* Clarendon Press, Oxford.

CHAPTER 3

1. Based on M. Vertregt (1960). *Elements of astronautics,* Elsevier, Amsterdam, p. 84.

CHAPTER 4

1. J.E. Gunn and B.M. Tinsley (1975). An accelerating universe. *Nature, Lond.* **257**, 454.

2. W.H. McCrea (1971). The cosmical constant. *Q. Jl R. astr. Soc.* **12**, 240.

3. E.A. Milne (1934). A Newtonian universe. *Q. Jl Math.* **5**, 64.

4. W.H. McCrea and E.A. Milne (1934). Newtonian universes and the curvature of space. *Q. Jl Math.* **5**, 73.

5. P.T. Landsberg (1973). Derivation of the differential equation for the simpler cosmological models. *Nature, Phys. Sci.* **242**, 104.

6. The parameter Q was introduced by E.R. Harrison (1976). Observational tests in cosmology. *Nature, Lond.* **260**, 591.

7. Problems (4.3) and (4.4) are based on P.T. Landsberg (1976). Q in cosmology. *Nature, Lond.* **263**, 217.

CHAPTER 5

1. P.T. Landsberg (1973). Deduction of the inverse square law from Newtonian cosmology. *Nature, Phys. Sci.* **244**, 66.

CHAPTER 6

1. G.A. Tammann (1974). The Hubble constant and the decleration parameter. In *Confrontation of cosmological theories with observational data* (Ed. M.S. Longair). I.A.U. Symposium No. 63, Reidel, Dortrecht.

2. P.T. Landsberg and N.T. Bishop (1975). A principle of impotence allowing for Newtonian cosmologies with a time-dependent gravitational constant. *Mon. Not. R. astr. Soc.* **171**, 279.

3. P.A.M. Dirac (1937). The cosmological constants. *Nature, Lond.* **139**, 323.

4. T.C. van Flandern (1975). A determination of the rate of change of G. *Mon. Not. R. astr. Soc.* **170**, 333.

5. D.S. Dearborn and D.N. Schramm (1974). Limits on the variation of G from clusters of galaxies. *Nature, Lond.* **247**, 441.

6. E. Böhm-Vitense and P. Szkody (1973). The interpretation of the two-color and color-magnitude diagrams of M15 and M92. *Astrophys. J.* **184**, 211.

7. W.A. Fowler (1972). In *Cosmology, fusion and other matters* (Ed. F. Reines), Adam Hilger, London.

8. R.V. Wagoner (1973). Big-bang nucleosynthesis revisited. *Astrophys. J.* **179**, 343.

CHAPTER 7

1. N.S. Kardashev (1967). Lemaître's universe and observations. *Astrophys. J.* **150**, L135.

2. G.C. McVittie and R. Stabell (1967). Lemaître universes and the log N-log S relation. *Astrophys. J.* **150**, L141.

CHAPTER 8

1. R.A. Lyttelton and H. Bondi (1959). On the physical consequences of a general excess of charge. *Proc. R. Soc.* **A252**, 313, equation (30).

2. H. Bondi and T. Gold (1948). The steady state theory of the expanding universe. *Mon. Not. R. astr. Soc.* **108**, 252.

3. A.M. Hillas and T.E. Cranshaw (1959, 1960). A comparison of the charges of electron, proton and neutron. *Nature, Lond.* **184**, 892 and **186**, 459.
   A more recent experiment, using a molecular beam deflection method, is reported by J.C. Zorn, G.E. Chamberlain, and V.W. Hughes (1963). Experimental limits for the electron-proton charge difference and for the charge of the neutron. *Phys. Rev.* **129**, 2566. For additional references and an experiment with an isolated macroscopic body, see R.W. Stover, T.I. Moran, and J.W. Trischka (1967). Search for electron-proton charge inequality by charge measurements on an isolated macroscopic body. *Phys. Rev.* **164**, 1599.

CHAPTER 9

1. See, for example, G.C. McVittie (1965). *General relativity and cosmology*, p. 143, Chapman and Hall, London. That the models are also non-rotating will not be stated explicitly.

2. This concept and its relation to reversible changes are discussed in P.T. Landsberg (1961). *Thermodynamics with quantum statistical illustrations*, Interscience, New York.

3. P.T. Landsberg and D. Park (1975). Entropy in an oscillating universe. *Proc. R. Soc.* **A346**, 485; G. Neugebauer and W. Meier (1976). Friedman-Kosmen mit irreversiblen Expansionsverhalten. *Ann. Phys.* **33**, 161; J. Pachner (1965). An oscillating isotropic universe without singularity. *Mon. Not. R. astr. Soc.* **131**, 173.

4. See, for example, G.L. Murphy (1973). Big-bang model without singularities. *Phys. Rev.* **8**, 4231.

5. M.J. Rees, R. Ruffini, and J.A. Wheeler (1974). *Black holes, gravitational waves and cosmology*, Gordon and Breach, New York.

6. See P.T. Landsberg, op. cit.

7. For further details see Y.K. Huang (1972). A special class of ideal quantum gases. *Am. J. Phys.* **40**, 1261; P.T. Landsberg (1974). The equations of state of some ideal fluids. *J. Phys.* **A7**, 859.

CHAPTER 10

1. J.A. Terrell (1975). Radio galaxies and local quasars. *Nature, Lond.* **258**, 132.

2. P.T. Landsberg and D.A. Evans (1972). *What Olbers might have said* in *The emerging universe* (Eds. W.C. Saslaw and K.C. Jacobs), University Press of Virginia.

3. For an account of general relativistic cosmology, see, for example, S. Weinberg (1972). *Gravitation and cosmology*, Wiley, New York, especially Chapter 14.

4. See Weinberg, op. cit., p. 410.

5. See Weinberg, op. cit., p. 422.

6. See Weinberg, op. cit., pp. 418–20.

7. See Weinberg, op. cit., p. 415.

CHAPTER 11

1. S. Weinberg (1972). *Gravitation and cosmology*, Wiley, New York, p. 459.

2. The general relativistic calculation is discussed in Weinberg, op. cit., pp. 451ff.

3. This is a special case of Weinberg, op. cit., eqn (14.71).

4. For a list of source counts, see Weinberg, op. cit., p. 458.

5. J.P. Loys de Cheseaux (1744). Sur la force de la lumière et sa propagation dans l'ether in *Traite de la comete qui a paru en Decembre 1743*, M.M. Bousquet et Cie., Geneva, pp. 223–9.

6. W. Olbers (1826). Über die Durchsichtigkeit des Weltraumes. *Astronom. Jahrbuch fur 1826*, pp. 110–21.

7. A historical discussion is given in S.L. Jaki (1967). Olbers', Halley's or whose paradox? *Am. J. Phys.* **35**, 200.

8. J.R. Gott III, J.E. Gunn, D.N. Schramm, and B.M. Tinsley (1974). An unbound universe? *Astrophys. J.* **194**, 543.

9. A.R. Sandage (1973). The red-shift–distance relation VII. *Astrophys. J.* **183**, 743.

10. J.E. Gunn and J.B. Oke (1975). Spectrophotometry of faint cluster galaxies and the Hubble diagram: an approach to cosmology. *Astrophys. J.* **195**, 255.

11. J.E. Gunn and B.M. Tinsley (1975). An accelerating universe. *Nature, Lond.* **257**, 454.

12. See Weinberg, op. cit., p. 489.

13. A.R. Sandage (1961). The ability of the 200-inch telescope to discriminate between selected world models. *Astrophys. J.* **133**, 355.

14. W.A. Baum (1972). The diameter–red-shift relation. In *External galaxies and quasi-stellar objects* (Ed. D.S. Evans), I.A.U. Symposium No. 44, Reidel, Dortrecht.

APPENDIX B

1. P.T. Landsberg and N.T. Bishop (1975). A cosmological deduction of the order of magnitude of an elementary-particle mass and of the cosmological coincidences. *Phys. Lett.* **53A**, 109.

2. P.A.M. Dirac (1974). Cosmological models and the large number hypothesis. *Proc. R. Soc.* **A338**, 439.

3. V.P. Chechev and Y.M. Kramarovsky (1972). Are the fundamental constants constant? *Contemp. phys.* **13**, 61.

4. P.S. Wesson (1973). The implications for geophysics of modern cosmologies in which G is variable. *Q. Jl R. astr. Soc.* **14**, 9.

5. J.K. Lawrence and G. Szamosi (1974). Statistical physics, particle

masses and the cosmological coincidences. *Nature, Lond.* **252**, 538.

APPENDIX C

1. P.T. Landsberg and B.M. Brown (1973). Limits on the cosmological constant and space curvature in Friedmann cosmologies. *Astrophys. J.* **182**, 653.

2. P.T. Landsberg and R.K. Pathria (1974). Cosmological parameters for a restricted class of closed big-bang universes. *Astrophys. J.* **192**, 577.

3. R. Stabell and S. Refsdal (1966). Classification of general relativistic world models. *Mon. Not. R. astr. Soc.* **132**, 379.

4. W. Rindler (1969). Limits on the cosmological declaration parameter. *Astrophys. J.* **157**, *L*. 147.

5. See, for example, R.K. Pathria (1972). The universe as a black hole. *Nature, Lond.* **240**, 298.

6. G.O. Abell (1965). Clustering of galaxies. *A. Rev. Astr. and Astrophys.* **3**, 1.

AUTHOR INDEX

SUBJECT INDEX

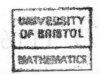